MATERIALS SCIENCE AND TECHNOLOGIES

ULTRA-HIGH PERFORMANCE CONCRETE

RESEARCH AND DEVELOPMENTS

Materials Science and Technologies

Additional books in this series can be found on Nova's website under the Series tab.

Additional E-books in this series can be found on Nova's website under the E-book tab.

Transportation Infrastructure - Roads, Highways, Bridges, Airports and Mass Transit

Additional books in this series can be found on Nova's website under the Series tab.

Additional E-books in this series can be found on Nova's website under the E-book tab.

MATERIALS SCIENCE AND TECHNOLOGIES

ULTRA-HIGH PERFORMANCE CONCRETE

RESEARCH AND DEVELOPMENTS

GARY M. AVALOS
EDITOR

nova publishers
New York

For permission to use material from this book please contact us:
Telephone 631-231-7269; Fax 631-231-8175
Web Site: http://www.novapublishers.com

Library of Congress Cataloging-in-Publication Data

ISBN: 978-1-62948-461-7

Published by Nova Science Publishers, Inc. † New York

Contents

PREFACE

This book documents the research, development, and deployment of UHPC components and includes information about its materials and production, mechanical properties, structural design and structural testing, durability and durability testing, and actual and potential applications. The book concludes with recommendations for the future direction for UHPC applications in the United States.

Chapter 1 – Advances in the science of concrete materials have led to the development of a new class of cementitious composites, namely ultra-high performance concrete (UHPC). The mechanical and durability properties of UHPC make it an ideal candidate for use in developing new solutions to pressing concerns about highway infrastructure deterioration, repair, and replacement. Since 2000, when UHPC became commercially available in the United States, a series of research projects has demonstrated the capabilities of the material. Three State transportation departments have deployed UHPC components within their infrastructure, and many more are actively considering the use of UHPC. This TechNote provides an introduction to UHPC and discusses practical considerations associated with it.

Chapter 2 – The term Ultra-High Performance Concrete (UHPC) refers to a relatively new class of advanced cementitious composite materials whose mechanical and durability properties far surpass those of conventional concrete. This class of concrete has been demonstrated to facilitate solutions that address specific problems in the U.S. highway bridge infrastructure. Initial material development research on UHPC began more than two decades ago. First structural deployments began in the late 1990s. First field deployments in the U.S. highway transportation infrastructure began in 2006. For this study, UHPC-class materials are defined as cementitious-based composite materials with discontinuous fiber reinforcement that exhibit compressive strength above 21.7 ksi (150 MPa), pre- and post-cracking tensile strength above 0.72 ksi (5 MPa), and enhanced durability via a discontinuous pore structure.

The report documents the state of the art with regard to the research, development, and deployment of UHPC components within the U.S. highway transportation infrastructure. More than 600 technical articles and reports covering research and applications using UHPC have been published in English in the last 20 years, with many more published in other languages. The report includes information about materials and production, mechanical properties, structural design and structural testing, durability and durability testing, and actual and potential applications. The report concludes with recommendations for the future direction for UHPC applications in the United States.

In: Ultra-High Performance Concrete
Editor: Gary M. Avalos

ISBN: 978-1-62948-461-7

Chapter 1

TECHNOTE: ULTRA-HIGH PERFORMANCE CONCRETE*

Ben Graybeal

INTRODUCTION

Advances in the science of concrete materials have led to the development of a new class of cementitious composites, namely ultra-high performance concrete (UHPC). The mechanical and durability properties of UHPC make it an ideal candidate for use in developing new solutions to pressing concerns about highway infrastructure deterioration, repair, and replacement.[1] Since 2000, when UHPC became commercially available in the United States, a series of research projects has demonstrated the capabilities of the material. Three State transportation departments have deployed UHPC components within their infrastructure, and many more are actively considering the use of UHPC. This TechNote provides an introduction to UHPC and discusses practical considerations associated with it.

DEFINITION

UHPC is a cementitious composite material composed of an optimized gradation of granular constituents, a water-to-cementitious materials ratio less than 0.25, and a high percentage of discontinuous internal fiber reinforcement. The mechanical properties of UHPC include compressive strength greater than 21.7 ksi (150 MPa) and sustained postcracking tensile strength greater than 0.72 ksi (5 MPa).† UHPC has a discontinuous pore structure that

---

* This publication, FHWA Publication No. FHWA-HRT-11-038, was published March 2011 by the U.S. Department of Transportation, Federal Highway Administration.

† The tensile behavior of UHPC may generally be defined as "strain-hardening," a broad term defining concretes wherein the sustained postcracking strength provided by the fiber reinforcement is greater than the cementitious matrix cracking strength. However, the definitional dependence on cementitious matrix cracking strength may in - appropriately include or exclude some concretes that exhibit dissimilar precracking and

reduces liquid ingress, significantly enhancing durability as compared to conventional and high-performance concretes.

APPLICATIONS

UHPC is being considered for use in a wide variety of highway infrastructure applications. The high compressive and tensile strengths allow for the redesign and optimization of structural elements. Concurrently, the enhanced durability properties facilitate a lengthening of design life and allow for potential use as thin overlays, claddings, or shells.

In the United States, UHPC has been used in three prestressed concrete girder simple-span bridges. The first two, located in Iowa and Virginia, used UHPC as a replacement for conventional concrete within I-girder shape members (see figure 1). In both cases, the tensile properties of UHPC were engaged to allow for the elimination of the mild steel reinforcement shear stirrups. The third bridge, located Iowa, used a prestressed deck-bulb-double-tee girder shape. This girder shape was optimized to engage the mechanical and durability properties of UHPC in a shape that facilitated accelerated construction.[2–4]

An optimized bridge redecking system has also been developed.[5] The two-way ribbed precast slab system, also known as a waffle slab, uses the mechanical and durability properties of UHPC to create a resilient, lightweight deck. This concept has been tested and is scheduled to be deployed by the Iowa Department of Transportation in 2011.[6]

Figure 1. This Wapello County, Iowa, structure was the first UHPC bridge constructed in the United States.

UHPCs have demonstrated exceptional performance when used as a field-cast closure pour or grout material in applications requiring the onsite connection of multiple prefabricated elements.[7] This use of UHPC has gained significant momentum recently, with States around the country considering the application. In 2009, two bridges using field-cast UHPC to create deck-level connections between precast concrete elements were constructed in New York. In one case, the UHPC was used in transverse connections between precast deck panels. In the other case, the UHPC was used in longitudinal connections between the

postcracking strength levels. The postcracking tensile strength and strain capacity of UHPC is highly dependent on the type, quantity, dispersion, and orientation of the internal fiber reinforcement.

top flanges of deck-bulb-tee girders (see figure 2). Field-cast UHPC will also be used in the longitudinal and transverse deck-level connections of the UHPC waffle slab bridge.

UHPC is also being investigated for use in a variety of other applications. These applications include precast concrete piles, seismic retrofit of substandard bridge substructures, thin-bonded overlays on deteriorated bridge decks, and security and blast mitigation applications. (See references 8–13.) In a general sense, UHPC has proven to be particularly relevant in applications where conventional solutions are lacking. For example, conventional connection solutions have hindered the use of prefabricated elements; field-cast UHPC allows for a redesign and simplification of the system while simultaneously promoting long-term durability.[7]

AVAILABILITY

The development of concretes within the UHPC class has progressed in recent years. The most readily available UHPC product in the United States is a proprietary product sold by a large multinational construction materials supplier. This product has undergone a signifi cant body of testing in order to demonstrate its specific characteristics. Alternate UHPC products are available in other parts of the world. Most notably, there are approximately five commercialized products available in Europe. Some of these product manufactures appear to be monitoring the U.S. market in preparation for the launch of competing UHPC products. Additionally, various research programs in Europe are facilitating the development of non-proprietary UHPC products produced from locally available constituents. Research programs in the United States are also progressing along this path.[14]

Source: New York State Department of Transportation.

Figure 2. Longitudinal connections are cast between deck-bulb-tee girders on the Route 31 Bridge in Lyons, NY. Field-cast UHPC can simplify connection details and ease constructability.

MIXING AND CASTING

UHPC is sufficiently similar to conventional concrete that the large majority of conventional concreting operations remain relevant and applicable. Nearly any conventional concrete mixer will mix UHPC. However, it must be recognized that UHPC requires increased energy input compared to conventional concrete, so mixing time will be increased. This increased energy input, in combination with the reduced or eliminated coarse aggregate and low water content, necessitates the use of modified procedures to ensure that the UHPC does not overheat during mixing. This concern can be addressed through the use of a high-energy mixer or by lowering the temperatures of the constituents and partially or fully replacing the mix water with ice. These procedures have allowed UHPC to be mixed in conventional pan and drum mixers, including ready-mix trucks (see figure 3).

The placement of UHPC may immediately follow mixing or be delayed while additional mixes are completed. Although the dwell time prior to the initiation of the cement hydration reactions can be influenced by factors such as temperature and chemical accelerators, it frequently requires multiple hours before UHPC will begin to set. During any extended dwell time, the UHPC should not be allowed to self-desiccate.

Casting of fiber-reinforced concretes requires special con siderations in terms of placement operations. UHPCs tend to exhibit rheological behaviors similar to conventional self-consolidating concretes, thus possibly necessitating additional form preparation but also allowing for reduced during-cast efforts. Internal vibration of UHPC is not recommended due to the fiber reinforcement, but limited external form vibration can be engaged as a means to facilitate the release of entrapped air.

Figure 3. UHPC is delivered via truck chute to a precast girder form. UHPC can be used in both precast and field-cast applications.

The long-term mechanical and durability properties of UHPC can be affected by casting procedures because the dispersion and orientation of the fiber reinforcement is influenced by the casting. First, fiber reinforcement tends to show a preference for aligning in the direction of flow during casting. This behavior must be recognized and considered when the casting sequence for a component is developed. Second, the ability of the fiber reinforcement to be maintained in suspension in the UHPC is dependent on the rheology of the concrete. Thus, any modification of the rheology or overreliance on form vibration must be carefully considered.

Curing Procedures

Applying appropriate curing methods is essential to the performance of any concrete, especially UHPC. Like all concretes, UHPCs require hydration water, but unlike other concretes, UHPCs have been engineered to require very little additional water, instead facilitating appropriate rheological behaviors through the use of an optimized gradation of granular materials. The reduced water content in a UHPC mix necessitates careful attention to curing practices so as not to allow the included water to escape prior to hydration.

Immediately after casting, any exposed UHPC surface needs to be sealed with an impermeable layer. Metal, plastic, or plastic-coated wood are appropriate materials with which to seal the surface. The seal must rest against the UHPC and should not allow for any space between the covering material and the fresh concrete. Sealing of the surface eliminates the possibility of surface dehydration, which can lead to cracking and significant degradation of final material properties.

Supplemental heat may be applied to UHPC castings after placement in order to accelerate setting behaviors and attainment of final properties. It is important to ensure that any added heat serves to raise the temp erature of the UHPC while not allowing material dehydration.

The UHPC should remain sealed in the formwork until it has attained sufficient properties to allow it to self-support and not self-desiccate. A compressive strength of 14 ksi (97 MPa) is frequently used as a surrogate value to indicate the attainment of an acceptable level of hydration.

It is possible to supplement the natural curing process of UHPC through the use of a steam treatment. This treatment can both enhance the final mechanical and durability properties of UHPC and accelerate the acquisition of said properties. A common steam treatment consists of subjecting the UHPC to a 194 °F (90 °C), 95 percent humidity environment for at least 2 days. If applied, this treatment frequently occurs at a precast concrete plant soon after form stripping. This treatment is not necessary and can be ignored if the properties of the as-cast UHPC are appropriate for the application being considered.

Testing Procedures

In general, well-established testing procedures for conventional concrete are applicable to UHPC. However, in some instances, procedures may need to be modified to appropriately capture the true behaviors of the UHPC.[15–17] Compression testing is a prime example. The conventional test method is generally appropriate, but compressive strengths as high as 35 ksi (240 MPa) may necessitate smaller specimen sizes, different specimen shapes, higher test machine capacities, or different specimen preparation techniques.

Flow Testing

Mix quality of self-consolidating concretes is frequently assessed through a slump cone flow test. With UHPC, it is more common to use the ASTM C1437 test, which measures flow of hydraulic cement mortars.[18] Both an initial flow reading and a dynamic flow reading are recorded. Frequently, this test is completed immediately after mixing to assess consistency between mixes and appropriateness for casting.

Figure 4. Either cylinders or cubes can be used for compression testing of UHPC.

Compression Testing

Compressive strength is arguably the most readily captured and used property of concrete. Research has demonstrated that standard concrete compression testing methods (i.e., ASTM C39, ASTM C109) are applicable to UHPC.[19,20] However, these test methods may benefit from slight modification to facilitate efficient use. Most notably, a loading rate of 150 psi/s (1 MPa/s) has been demonstrated to be acceptable, thus allowing for individual tests to be completed in a reasonable timeframe.[15]

Additional research has demonstrated that, at the high strength levels achieved with UHPC, cube compres sive tests are an appropriate substitute for cylinder compression tests (see figure [4,21,22] Companion cylinder and cube strength results tend to be within 5 percent of one another, thus allowing for direct substitution of results. If cylinder tests are used for specimens at strength levels above those appropriate for capping methods, both cylinder ends must be ground planar to within the ASTM C39 specification of 0.5 degrees. Also recognize that the high compressive strengths of UHPC may necessitate the use of higher capacity compression testing platens and machines.

Modulus of Elasticity Testing

UHPC does not present any specific challenges or require any specific modifications to the standard ASTM C469 test method for static modulus of elasticity.[23]

Tensile Testing

UHPC in particular and fiber-reinforced concrete in general present specific challenges in terms of quantifying tensile behaviors. Tensile cracking strength of UHPC can be measured through the same tests used for conventional concrete. However, special precautions must be taken to ensure that the true cracking strength, not a postcracking fiber reinforcement-enhanced strength, is recorded as the cracking strength. Prism flexure testing and split cylinder testing are both appropriate means of determining first cracking, but the specimen must be closely monitored to capture the load at first cracking, since the load may continue to increase thereafter without noticeable change in global specimen behavior. Monitoring of specimens can be done visually, audibly, or through the use of nondestructive testing equipment.[15,24]

The postcracking tensile behavior of UHPC is one of the unique properties that differentiate it from conventional concrete. UHPC generally falls into the category of strain-hardening fiber-reinforced concrete, which means that the postcracking strength provided by the fiber reinforcement bridging a crack is equal to or greater than the cracking strength of the cementitious matrix. This behavior is responsible for the multicracking response of UHPC components and allows for the potential inclusion of UHPC tensile strength and strain capacities in structural design calculations.

No standardized test for quantitative determination of the full range of UHPC tensile behaviors exists in the United States. Overseas, most notably in France, a set of standardized prism flexure tests is used to quantify the tensile response. Development of standardized tests is underway in the United States, with an emphasis on the development of a test to directly capture the tensile response by pulling cast or extracted UHPC prismatic specimens (see figure 5). Until such a test is available, the use of non-standardized tests or the derivation of tensile response from the testing of a full-scale structural component may be necessary.

Chloride Penetration Testing

Conventional, ponding-type chloride penetration tests, such as AASHTO T259, can be completed on UHPC specimens.[25] Exposed steel fiber reinforcement may corrode during long-duration tests, but this should not impact the overall test results. When completing such a test, recognize that UHPC frequently contains un hydrated cementitious constituents. Thus, initial water penetration can lead to additional hydration and a further reduction in permeability. Also recognize that testing extracted powder samples for chloride concentration may require extra processing in order to remove included fiber pieces.

Rapid chloride penetrability testing can be completed on UHPC samples as well. Whether reinforced by steel or organic fibers, the reinforcement within the UHPC matrix is generally dispersed and discontinuous. Tests have shown that steel fiber reinforcement within

UHPC does not provide a direct path to complete an electric circuit. As such, completing the ASTM C1202 test on a UHPC cylinder containing 0.5-inch (13-mm)- long steel fiber reinforcement can provide a comparative result indicative of chloride ion penetrability.[26]

Freeze-Thaw Durability Tests

Conventional freeze-thaw test methods, such as ASTM C512, can be applied to UHPC.[27] However, the un hydrated cementitious particles frequently present in UHPC can hydrate when contacted by water. Thus, any exposure of the UHPC to liquid water can result in surface penetration of the water, localized hydration, and increased dynamic modulus. In practice, freeze-thaw testing can indi cate that the UHPC performance is bolstered through exposure to these conditions, while in actuality the thaw portion of the cycle is facilitating delayed hydration and a requisite dynamic modulus increase.

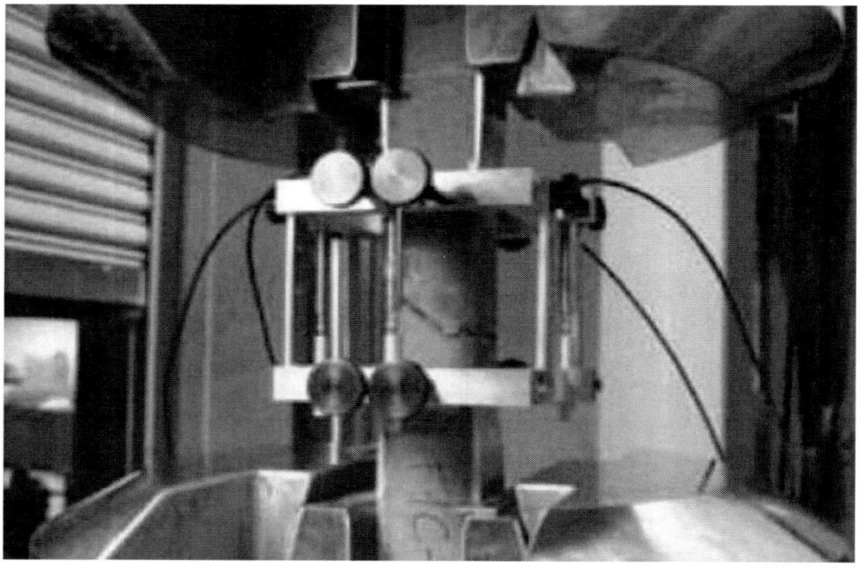

Figure 5. Testing procedures aimed at directly capturing the tensile behavior of UHPC are under development.

Scaling and Other Durability Tests

Other conventional concrete durability test methods can generally be applied to UHPC specimens. Many of these tests can provide comparative results indicating the relative durability of UHPC in terms of conventional concrete. However, many of these tests use subjective, qualitative measures to assess performance. Since these measures have been developed for use with conventional concrete, UHPC may exceed the anticipated performance range, thus making comparisons between individual UHPCs difficult.

SAMPLE PREPARATION AND EXTRACTION

The creation and/or acquisition of UHPC samples for material testing does not differ significantly from methods used for conventional concrete. Cast specimens may be fabricated into any shape desired through the use of conventional concrete molds. However, it is important to recognize that UHPC flow during casting can cause preferential fiber orientation that may impact later test results.

Extraction of specimens from larger components may be completed through methods normally used for conventional concrete. In general, UHPC and conven tional concrete are both composed of similar constituent materials. Not surprisingly, conventional cutting and grinding equipment has been found to be both applicable and effective.

STRUCTURAL DESIGN, ANALYSIS, AND MODELING

As with any structural material whose physical properties are outside the bounds of existing structural design specifications, the design of UHPC structural components presents challenges not normally present in the design of routine reinforced concrete components. However, it must be recognized that the properties of UHPC can be determined and that full-scale structural testing has demonstrated the fundamental structural performance of UHPC. (See references 2, 3, and 28–32.) The understanding and appropriate application of basic structural engineering principles such as flexure and shear are critical in the development of effective UHPC designs.

Analysis of UHPC structural components is not necessarily more complex than analysis of conventional reinforced concrete structures. However, it is imperative that the analysis be completed rationally without allowing preconceived notions of reinforced concrete behavior to cloud the computations. For example, the compressive stress-strain response of conventional concrete has a parabolic shape that is sometimes modeled through a rectangular stress block. Such a stress block would only be appropriate for UHPC if appropriate factors were applied to adjust the block to match the stiffer and more linear UHPC response. Moreover, it is frequently desirable to include the sustained tensile capacity of UHPC in an analysis. Although this extra stress block may increase the complexity of the calculation, it does not introduce new theory and should be able to be computed appropriately.

More complex analytical procedures for UHPC have been developed recently.[33] These procedures rely on commercially available finite element software to accurately model the structural response of UHPC components and structures (see figure 6).

INSPECTION

Inspection of structural concrete components in the deployed infrastructure involves a set of well-known processes and inspection tools. Inspection of UHPC is no different than that of conventional concrete, except that the years of experience associated with UHPC assessment has not yet been developed. In general, any inspection procedure associated with conventional concrete should be applicable to UHPC structures. The simplest inspection process, namely visual inspection, will differ primarily in that the scale of defects considered significant will be reduced. UHPC tends to exhibit tightly spaced, small-width cracks that are difficult to locate with the unaided eye.[34] The use of an evaporative, penetrating liquid, such as alcohol, can greatly simplify the identification of cracks.

Advanced inspection technologies, such as acoustic methodologies, can be simplified and expanded when used on UHPC structures. The comparatively homogenous composition of UHPC, in conjunction with its increased modulus of elasticity, expands the potential technologies applicable to this concrete to include some items previously reserved for metals.[35]

Conclusion

Whether used to facilitate accelerated construction, lengthen span ranges, or rehabilitate substandard infrastructure, UHPC can facilitate the development of unique solutions to existing challenges. As with any new material, utilization will grow as innovative applications are developed and market demand intensifies. A decade of research and deployment efforts by groups associated with the U.S. highway transportation sector has demonstrated that UHPC is a material both capable of and poised for future deployment in infrastructure-scale applications.

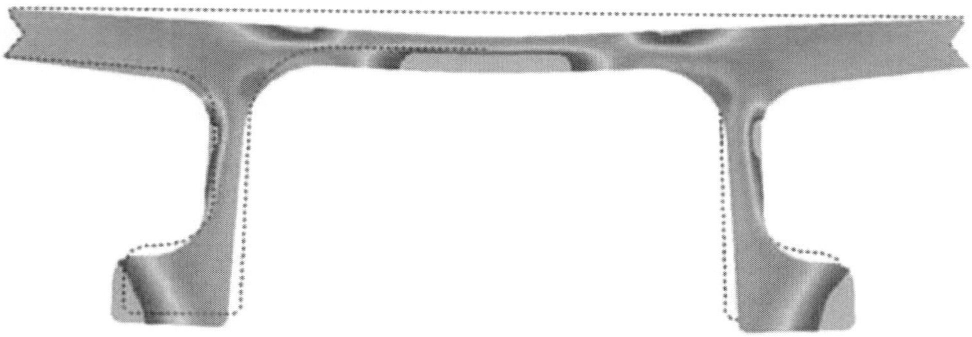

Figure 6. Finite element computer modeling techniques have been calibrated and demonstrated capable of modeling UHPC structural response.

References

[1] Graybeal, B. (2009). "UHPC Making Strides," *Public Roads*, Vol. 72, No. 4, pp. 17–21, Federal Highway Administration, McLean, VA.

[2] Graybeal, B. (2009). *Structural Behavior of a 2nd Generation Ultra-High Performance Concrete Pi-Girder*, Report No. PB2009-115496, National Technical Information Service, Springfield, VA.

[3] Graybeal, B. (2009). *Structural Behavior of a Prototype Ultra-High Performance Concrete Pi-Girder*, Report No. PB2009-115495, National Technical Information Service, Springfield, VA.

[4] Graybeal, B. (2010). "Design, Fabrication, and Testing of a 2nd Generation Ultra-High Performance Concrete Pi-Girder," *Proceedings, 3rd International fib Congress*, Washington, DC.

[5] Garcia, H. (2007). *Analysis of an Ultra-High Performance Concrete Two-Way Ribbed Bridge Deck Slab*, Report No. PB 2007-112112, National Technical Information Service, Springfield, VA.

[6] Aaleti, S., Sritharan, S., Bierwagen, D., and Wipf, T. (2011). "Experimental Evaluation of Structural Behavior of Precast UHPC Waffle Bridge Deck Panels and Connections," *Proceedings, Transportation Research Board Annual Meeting*, Washington, DC.

[7] Graybeal, B. (2010). *Behavior of Field-Cast Ultra-High Performance Concrete Bridge Deck Connections Under Cyclic and Static Structural Loading*, Report No. PB2011-101995, National Technical Information Service, Springfield, VA.

[8] Vande Voort, T., Suleiman, M., and Sritharan, S. (2008). *Design and Performance Verification of UHPC Piles for Deep Foundations*, Final Report, Iowa Highway Research Board Project TR-558, Iowa State University, Ames, IA.

[9] Massicotte, B. and Boucher-Proulx, G. (2010). "Seismic Retrofitting of Bridge Piers with UHPFRC Jackets," *Designing and Building with UHPFRC: State of the Art and Development*, pp. 531–540 , Wiley-ISTE, London.

[10] Brühwiler, E., and Denarié, E. (2008). "Re habil tation of Concrete Structures Using Ultra-High Performance Fiber-Reinforced Concrete," *Proceedings, The Second International Symposium on Ultra-High Performance Concrete*, Kassel, Germany.

[11] Schmidt, C., Riedl, S., Geisenhanslüke, C., and Schmidt, M. (2008). "Strengthening and Rehabilitation of Pavements Applying Thin Layers of Reinforced Ultra-High Performance Concrete (UHPC-White Topping)," *Proceedings, The Second Inter national Symposium on Ultra-High Performance Concrete*, Kassel, Germany.

[12] Rebentrost, M., and Wight, G. (2009). "Investigation of UHPFRC Slabs Under Blast Loads," *Proceedings, Ultra-High Performance Fiber Reinforced Concrete 2009*, Marseille, France.

[13] Green, B. (2010). "An Investigation of UHPC/RPC Materials for Enhanced Penetration Resistance," Presented at the American Concrete Institute Fall Convention, Pittsburgh, PA.

[14] Wille, K., Naaman, A., and Parra-Montesinos, G. (2011). "Ultra-High Performance Concrete with Compressive Strength Exceeding 150 MPa (22 ksi): A Simpler Way," *ACI Materials Journal*, Vol. 108, No. 1, pp. 46–54, American Concrete Institute, Farmington Hills, MI.

[15] Graybeal, B. (2006). *Material Property Character ization of Ultra-High Performance Concrete*, Report No. FHWA-HRT-06-103, Federal Highway Administration, McLean, VA.

[16] Graybeal, B. and Tanesi, J. (2007). "Durability of an Ultrahigh-Performance Concrete," *ASCE Journal of Materials in Civil Engineering*, Vol. 19, No. 10, pp. 850–854, American Society of Civil Engineers, Reston, VA.

[17] Graybeal, B. (2010). *Simultaneous Structural and Environmental Loading of an Ultra-High Performance Concrete Component*, Report No. PB2010-110331, National Technical Information Service, Springfield, VA.

[18] ASTM International (2007). "Standard Test Method for Flow of Hydraulic Cement Mortar," *ASTM Standard C1437*, West Conshohocken, PA.

[19] ASTM International (2010). "Standard Test Method for Compressive Strength of Cylindrical Concrete Specimens," *ASTM Standard C39*, West Conshohocken, PA.

[20] ASTM International (2008). "Standard Test Method for Compressive Strength of Hydraulic Cement Mortars (Using 2-in. or [50-mm] Cube Specimens)," *ASTM Standard C109*, West Conshohocken, PA.

[21] Graybeal, B. and Davis, M. (2008). "Cylinder or Cube: Strength Testing of 80 to 200 MPa (11.6 to 29 ksi) Ultra-High-Performance Fiber-Reinforced Concrete," *ACI Materials Journal*, Vol. 105, No. 6, pp. 603–609, American Concrete Institute, Farmington Hills, MI.

[22] Graybeal, B. (2007). "Compressive Behavior of Ultra-High Performance Fiber-Reinforced Concrete," *ACI Materials Journal*, Vol. 104, No. 2, pp. 146–152, Farmington Hills, MI.

[23] ASTM International (2010). "Standard Test Method for Static Modulus of Elasticity and Poisson's Ratio of Concrete in Compression," *ASTM Standard C469*, West Conshohocken, PA.

[24] Graybeal, B. (2006). "Practical Means for Determination of the Tensile Behavior of Ultra-High Performance Concrete," *Journal of ASTM International*, Vol. 3, No. 8, ASTM International, West Conshohocken, PA.

[25] American Association of State Highway and Transportation Officials (2006). "Standard Method of Test for Resistance of Concrete to Chloride Ion Penetration," AASHTO Standard T259, *Standard Specifications for Transportation Materials and Methods of Sampling and Testing*, Washington, DC.

[26] ASTM International (2010), "Standard Test Method for Electrical Indication of Concrete's Ability to Resist Chloride Ion Penetration," *ASTM Standard C1202*, West Conshohocken, PA.

[27] ASTM International (2008). "Standard Test Method for Resistance of Concrete to Rapid Freezing and Thawing," *ASTM Standard C666*, West Conshohocken, PA.

[28] Graybeal, B. (2006). *Structural Behavior of Ultra-High Performance Concrete Prestressed I-Girders*, Report No. FHWA-HRT-06-115, Federal Highway Administration, McLean, VA.

[29] Graybeal, B. (2008). "Flexural Behavior of an Ultrahigh-Performance Concrete I-Girder," *ASCE Journal of Bridge Engineering*, Vol. 13, No. 6, pp. 602–610, American Society of Civil Engineers, Reston, VA.

[30] Crane, C.K. (2010). "Shear and Shear Friction of Ultra-High Performance Concrete Bridge Girders," Ph.D. Dissertation, Georgia Institute of Technology, Atlanta, GA.

[31] Harris, D. (2004). "Characterization of the Punching Shear Capacity of Thin UHPC Plates," Master's Thesis, Virginia Polytechnic Institute and State University, Blacksburg, VA.

[32] Degen, B. (2006). "Shear Design and Behavior of Ultra-High Performance Concrete," Master's Thesis, Iowa State University, Ames, IA.

[33] Chen, L., and Graybeal, B. (2010). *Finite Element Analysis of Ultra-High Performance Concrete: Modeling Structural Performance of an AASHTO Type II Girder and a 2nd Generation Pi-Girder*, Report No. PB2011-100864, National Technical Information Service, Springfield, VA.

[34] Meade, T., and Graybeal, B. (2010). "Flexural Response of Lightly Reinforced Ultra-High Performance Concrete Beams," *Proceedings, 3rd International fib Congress*, Washington, DC.

[35] Washer, G., P. Fuchs, B. Graybeal, and J. Hartmann. (2004). "Ultrasonic Testing of Reactive Powder Concrete," *IEEE Transactions on Ultrasonics, Ferroelectrics, and Frequency Control*, Vol. 51, No. 2, IEEE Ultrasonics, Ferroelectrics, and Frequency Control Society, New York, NY.

In: Ultra-High Performance Concrete
Editor: Gary M. Avalos

ISBN: 978-1-62948-461-7

Chapter 2

ULTRA-HIGH PERFORMANCE CONCRETE: A STATE-OF-THE-ART REPORT FOR THE BRIDGE COMMUNITY*

Henry G. Russell and Benjamin A. Graybeal

ABSTRACT

The term Ultra-High Performance Concrete (UHPC) refers to a relatively new class of advanced cementitious composite materials whose mechanical and durability properties far surpass those of conventional concrete. This class of concrete has been demonstrated to facilitate solutions that address specific problems in the U.S. highway bridge infrastructure. Initial material development research on UHPC began more than two decades ago. First structural deployments began in the late 1990s. First field deployments in the U.S. highway transportation infrastructure began in 2006. For this study, UHPC-class materials are defined as cementitious-based composite materials with discontinuous fiber reinforcement that exhibit compressive strength above 21.7 ksi (150 MPa), pre- and post-cracking tensile strength above 0.72 ksi (5 MPa), and enhanced durability via a discontinuous pore structure.

The report documents the state of the art with regard to the research, development, and deployment of UHPC components within the U.S. highway transportation infrastructure. More than 600 technical articles and reports covering research and applications using UHPC have been published in English in the last 20 years, with many more published in other languages. The report includes information about materials and production, mechanical properties, structural design and structural testing, durability and durability testing, and actual and potential applications. The report concludes with recommendations for the future direction for UHPC applications in the United States.

* This is an edited, reformatted and augmented version of U.S. Department of Transportation, Federal Highway Administration, Publication No. . FHWA-HRT-13-060, dated June 2013.

FOREWORD

Ultra-high performance concrete (UHPC) is an advanced construction material that affords new opportunities for the future of the highway infrastructure. The Federal Highway Administration has been engaged in research on the optimal uses of UHPC in the highway bridge infrastructure since 2001 through its Bridge of the Future initiative. This report presents the state of the art in UHPC with regard to uses in the highway transportation infrastructure. Compiled from hundreds of references representing research, development, and deployment efforts around the world, this report provides a framework for gaining a deeper understanding of UHPC as well as a platform from which to increase the use of this class of advanced cementitious composite materials. This report will assist stakeholders, including State transportation departments, researchers, and design consultants, to grasp the capabilities of UHPC and thus use the material to address pressing needs in the highway transportation infrastructure.

Jorge Pagán-Ortiz
Director,
Office of Infrastructure
Research and Development

SI* (MODERN METRIC) CONVERSION FACTORS				
APPROXIMATE CONVERSIONS TO SI UNITS				
Symbol	When You Know	Multiply By	To Find	Symbol
		LENGTH		
in	inches	25.4	millimeters	mm
ft	feet	0.305	meters	m
yd	yards	0.914	meters	m
mi	miles	1.61	kilometers	km
		AREA		
in²	square inches	645.2	square millimeters	mm²
ft²	square feet	0.093	square meters	m²
yd²	square yard	0.836	square meters	m²
ac	acres	0.405	hectares	ha
mi²	square miles	2.59	square kilometers	km²
		VOLUME		
fl oz	fluid ounces	29.57	milliliters	mL
gal	gallons	3.785	liters	L
ft³	cubic feet	0.028	cubic meters	m³
yd³	cubic yards	0.765	cubic meters	m³
		NOTE: volumes greater than 1000 L shall be shown in m³		
		MASS		
oz	ounces	28.35	grams	g
lb	pounds	0.454	kilograms	kg
T	short tons (2000 lb)	0.907	megagrams (or "metric ton")	Mg (or "t")
		TEMPERATURE (exact degrees)		
°F	Fahrenheit	5 (F-32)/9 or (F-32)/1.8	Celsius	°C
		ILLUMINATION		
fc	foot-candles	10.76	lux	lx
fl	foot-Lamberts	3.426	candela/m²	cd/m²
		FORCE and PRESSURE or STRESS		
lbf	poundforce	4.45	newtons	N
lbf/in²	poundforce per square inch	6.89	kilopascals	kPa

APPROXIMATE CONVERSIONS FROM SI UNITS				
Symbol	When You Know	Multiply By	To Find	Symbol
		LENGTH		
mm	millimeters	0.039	inches	in
m	meters	3.28	feet	ft
m	meters	1.09	yards	yd
km	kilometers	0.621	miles	mi
		AREA		
mm^2	square millimeters	0.0016	square inches	in^2
m^2	square meters	10.764	square feet	ft^2
m^2	square meters	1.195	square yards	yd^2
ha	hectares	2.47	acres	ac
km^2	square kilometers	0.386	square miles	mi^2
		VOLUME		
mL	milliliters	0.034	fluid ounces	fl oz
L	liters	0.264	gallons	gal
m^3	cubic meters	35.314	cubic feet	ft^3
m^3	cubic meters	1.307	cubic yards	yd^3
		MASS		
g	grams	0.035	ounces	oz
kg	kilograms	2.202	pounds	lb
Mg (or "t")	megagrams (or "metric ton")	1.103	short tons (2000 lb)	T
		TEMPERATURE (exact degrees)		
°C	Celsius	1.8C+32	Fahrenheit	°F
		ILLUMINATION		
lx	lux	0.0929	foot-candles	fc
cd/m^2	$candela/m^2$	0.2919	foot-Lamberts	fl
		FORCE and PRESSURE or STRESS		
N	newtons	0.225	poundforce	lbf
kPa	kilopascals	0.145	poundforce per square inch	lbf/in^2

*SI is the symbol for th International System of Units. Appropriate rounding should be made to comply with Section 4 of ASTM E380. (Revised March 2003).

SECTION 1. INTRODUCTION

Background

Ultra-high performance concrete (UHPC) in its present form became commercially available in the United States in about 2000. [1] The Federal Highway Administration (FHWA) began investigating the use of UHPC for highway infrastructure in 2001 and has been working with State transportation departments to deploy the technology since 2002. This work has led to the use of UHPC in several bridge applications, including precast, prestressed girders; precast waffle panels for bridge decks; and as a jointing material between precast concrete deck panels and girders and between the flanges of adjacent girders. At the same time, research work has been underway at several universities in the United States.

In Canada, the first UHPC bridge was constructed in 1997. [2] This pedestrian bridge consists of a precast, post-tensioned space truss. At least 26 bridges have been built in Canada using UHPC in one or more components.

In Germany, a 12 million euro research program, begun in 2005, has just been completed. [3] That program, funded by the German Research Foundation, involved 34 research projects at more than 20 research institutes in Germany. The purpose of the program was to elaborate on the basic knowledge so that reliable technical standards could be developed. The goal was to make UHPC a reliable, commonly available, economically feasible, regularly applied material. Several bridges that use UHPC have been built in Germany.

In 2002, the first recommendations on the use of UHPC in structures were published in France. [4] This initial document addressed mechanical properties, structural design, and durability. Since 2002, several bridges have been built in France using UHPC. In 2009, several papers published in French recommended updates to the recommendations. [5,6,7] A similar set of design recommendations was developed for use in Japan. [8]

Other countries with bridges using UHPC include Australia, Austria, Croatia, Italy, Japan, Malaysia, the Netherlands, New Zealand, Slovenia, South Korea, and Switzerland. The literature search identified more than 90 completed bridges using UHPC in one or more components. A major research program is currently underway in South Korea to investigate the use of UHPC in cable-stayed bridges. [9] It is obvious, therefore, that UHPC is receiving worldwide attention.

The evolution of UHPC into its present formulation has been a gradual process occurring over many years. Several papers have summarized this development. Naaman and Wille identified many of the significant advances in the technology over the last 5 decades. [10] Buitelaar summarized the early developments in the Netherlands and Denmark. [11] Richard and Rossi described the developments in France. [12,13,14]

As part of its ongoing activities to implement UHPC in the United States, FHWA has requested a state-of-the-art report about the research, development, and deployment of UHPC. This document reports what has been done and looks ahead to what needs to be done to achieve appropriate applications in the U.S. highway infrastructure.

Objective

The objective of this report is to document the state of the art with regard to the research, development, and deployment of UHPC components within the U.S. highway transportation infrastructure. In addition, because much of the development and initial deployment of UHPC has occurred internationally, the report also documents work completed outside the U.S. highway sector. In addition, the report addresses what is needed to allow future wider implementation of UHPC.

Scope

Similar to how Graybeal defines it, this document defines UHPC-class materials as cementitiousbased composite materials with discontinuous fiber reinforcement, compressive strengths above 21.7 ksi (150 MPa), pre-and post-cracking tensile strengths above 0.72 ksi (5 MPa), and enhanced durability via their discontinuous pore structure. [1] However, the published literature does not always include sufficient information to determine whether the tested materials conformed to this definition. Unless an obvious reason existed to exclude an article, it is included in this report.

The authors identified more than 600 references relevant to this report. Some topics are described in more than one article by the same or similar combinations of authors. Some articles also provide updates on previous articles on the same topic. For this report, the most comprehensive documents readily available and written in English are used for most of the cited references. The other articles are listed in the Bibliography. Based on this approach, the majority of articles in this report come from the following publications:

- FHWA reports.
- Proceedings of the International Symposium on Ultra High Performance Concrete, Kassel, Germany, September 2004.
- Proceedings of the Second International Symposium on Ultra High Performance Concrete, Kassel, Germany, March 2008.
- Proceedings of the International Workshop on Ultra High Performance Fibre Reinforced Concrete—Designing and Building with UHPFRC: State of the Art Development, November 2009, Marseille, France.
- Proceedings of Hipermat 2012 3rd International Symposium on UHPC and Nanotechnology for High Performance Construction Materials, Kassel, Germany, March 2012.

The articles published in the proceedings of the international meetings are usually summaries of the research or applications and written in English. As such, the articles do not contain sufficient information for use in developing design guides or specifications. For more details, the articles often refer to a full report written in a language other than English. Use of these reports requires either a comprehension of the other language or an English translation.

Terminology

Various terms are used to refer to cementitious-based composite materials with high compressive strength and enhanced durability. These include the following:

- Compact reinforced composite (CRC).
- Densified small-particle (DSP) concrete.
- Fiber-reinforced high-performance concrete (FRHPC).
- High-performance fiber reinforced cement composite (HPFRCC).
- Macro defect free (MDF) concrete.
- Multi-scale fiber-reinforced concrete (MSFRC).
- Reactive powder concrete (RPC).
- Steel fibrous cement-based composite (SFCBC).
- Ultra-high performance concrete (UHPC).
- Ultra-high performance fiber-reinforced cementitious composite (UHPFRCC).
- Ultra-high performance fiber-reinforced concrete (UHPFRC).
- Ultra-high strength concrete (UHSC).
- Ultra-high strength cement-based composite.
- Ultra-high strength cementitious material.
- Ultra-high strength fiber-reinforced cementitious composite.

In addition, various patterns of hyphens are used to form compound adjectives. For this report, the product is generally called ultra-high performance concrete or UHPC unless it is necessary to differentiate the different types. Descriptions of some of the different types are provided by Rossi. [13] Although calling the different types by a single name may not be technically correct, it simplifies understanding the available information.

This report also refers to conventional concrete. Conventional concrete is composed of cementitious materials, fine and coarse aggregates, water, and admixtures. Compressive strengths are assumed to be in the range of 4 to 8 ksi (28 to 55 MPa).

Sustainability

Few articles have been published about the sustainability of UHPC compared with the number published about its material and engineering properties. Several authors have addressed the topic because of the increasing requirement to consider sustainability.

Racky determined that the energy and raw material consumption to produce a square reinforced column made of UHPC were 74 and 58 percent, respectively, of the quantities required for a Grade 40/50 (6/7 ksi) column. [15] He also pointed out that UHPC had greater frost and deicing salt resistance, a lower rate of carbonation, and greater chloride resistance than conventional concretes. Consequently, highway structures made with UHPC will have lower maintenance and repair costs in the future compared with conventional concrete bridges. However, sufficient data were lacking to perform realistic life cycle cost analyses.

Schmidt and Jerebic reported that the energy demanded for production of 1.3 yd^3 (1 m^3) of UHPC was approximately double that for conventional concrete. [16] However, when the total energy demand to construct the Gaertnerplatz bridge using UHPC and steel tubes was compared with the energy demand for an equivalent conventional prestressed concrete bridge, the increase was reduced to 25 percent. The largest component of energy was for the production of the steel tubes. If the steel tubes could be replaced by UHPC tubes, the estimated energy demand would drop by about 50 percent. When the CO_2 contributions from construction to the greenhouse effect were considered, the UHPC-steel combination had the largest value, the UHPC tube the lowest, and the prestressed concrete bridge was intermediate.

Sedran, Durand, and de Larrard reported that UHPC called Ceracem could be crushed and separated into sand and fibers. [17] The recycled sand could then be used as a replacement for river sand in self-leveling concrete with no loss of fluidity and no decrease in compressive strength.

Stengel and Schießl reported that the environmental impact of UHPC production was mainly caused by the production of the steel fibers, portland cement, and high-range water-reducing admixtures. [18] The effect of heat curing UHPC was not taken into account. In another study, life-cycle assessments of bridge structures were made using German standard Deutsches Institut für Normung (DIN) ISO 14040 ff. [19,20] The research concluded that the environmental impact of structures made with state-of-the-art UHPC was up to 2.5 times greater than with conventional concrete. The environmental impact could be decreased by reducing the amount of portland cement, steel fibers, and high-range water-reducing admixtures in the UHPC.

Costs

The initial unit quantity cost of UHPC far exceeds that of conventional concrete. Consequently, applications have focused on optimizing its use by reducing concrete member thickness, changing concrete structural shapes, or developing solutions that address shortcomings with existing non-concrete structural materials. As discussed in section 5, UHPC is a very durable product, and structures that use it are expected to have a longer service life and require less maintenance than structures built with conventional concrete.

Piotrowski and Schmidt conducted a life cycle cost analysis of two replacement methods for the Eder bridge in Felsberg, Germany. [21] One method used precast UHPC box girders filled with lightweight concrete. The second method used conventional prestressed concrete bridge members. Although the UHPC had higher initial costs, the authors predicted the life cycle cost over 100 years would be less for the UHPC bridge.

SECTION 2. MATERIALS AND PRODUCTION

Constituent Materials and Mix Proportions

UHPC formulations often consist of a combination of portland cement, fine sand, silica fume, high-range water-reducing admixture (HRWR), fibers (usually steel), and water. Small aggregates are sometimes used, as well as a variety of chemical admixtures. Different combinations of these materials may be used, depending on the application and supplier. Some of these are described in this section.

The UHPC used most often in North America for both research and applications is a commercial product known as Ductal®. Table 1 shows a typical composition of this material. [22]

Table 1. Typical composition of Ductal®

Material	lb/yd^3	kg/m^3	Percentage by Weight
Portland Cement	1,200	712	28.5
Fine Sand	1,720	1,020	40.8
Silica Fume	390	231	9.3
Ground Quartz	355	211	8.4
HRWR	51.8	30.7	1.2
Accelerator	50.5	30.0	1.2
Steel Fibers	263	156	6.2
Water	184	109	4.4

Aarup reported that CRC, developed by Aalborg Portland in 1986, consisted of large quantities of steel fibers (2 to 6 percent by volume), large quantities of silica fume, and a water-binder ratio of 0.16 or lower. [23]

The following recommendations for mix proportions were developed for use with commercially available constituent materials: [24]

- Cement with a moderate fineness and C_3A content significantly lower than 8 percent.
- Sand-to-cement ratio of 1.4 for a maximum grain size of 0.8 mm (0.03 inches).
- Silica fume with very low carbon content at 25 percent of the weight of cement.

- Glass powder with a median particle size of 67 x 10^{-6} inches (1.7 μm) at 25 percent of the weight of cement.
- High-range water-reducing admixture.
- Water-cement ratio of about 0.22.
- Steel fibers at 2.5 percent by volume.

By optimizing the cementitious matrix for compressive strength, packing density, and flowability; using very high strength, fine-diameter steel fibers; and tailoring the mechanical bond between the steel fiber and cement matrix, 28-day compressive strengths in excess of 30 ksi (200 MPa) on 2-inch (50-mm) cubes were achieved with no heat or pressure curing. [25] In addition, a tensile strength of 5.0 ksi (34.6 MPa) at a strain of 0.46 percent was obtained. The UHPC incorporated materials available in the United States and was mixed in a conventional concrete mixer. Table 2 gives one mix proportion.

Table 2. UHPC mix proportions of CRC by weight [25]

Material	Proportions
Portland Cement	1.0
Fine Sand[1]	0.92
Silica Fume	0.25
Glass Powder	0.25
HRWR	0.0108
Steel Fibers	0.22 to 0.31
Water	0.18 to 0.20

[1] Maximum size of 0.008 inches (0.2 mm).

Habel et al. reported that it is possible to produce self-consolidating UHPC for use in precast products and cast-in-place (CIP) applications without requiring heat or pressure treatment during curing. [26] This mix design was further developed and implemented in a research program conducted by Kazemi and Lubell. [27]

Holschemacher and Weißl investigated different mix proportions to minimize material costs without sacrificing the beneficial properties of UHPC. [28] Through careful selection of aggregates, cement type, cementitious materials, inert filler, and HRWR, it was possible to produce UHPC with good workability and moderate material costs.

The concept of combining different size molecular admixtures to facilitate UHPC dispersion was studied by Plank et al. [29]

The possibility of replacing silica fume in UHPC with metakaolin, pulverized fly ash, limestone microfiller, siliceous microfiller, micronized phonolith, or rice husk ash has been investigated. [30,31] The use of local materials rather than proprietary products has also been pursued. [32,33]

Schmidt et al. reported two mix proportions for a bridge in Germany. [34] The first mix contained 1,854 lb/yd^3 (1,100 kg/m^3) of cement, 26-percent silica fume as a percentage of the cement content, quartz sand, 6 percent steel fibers by volume, HRWR, and a water-binder ratio of 0.14. The second mix contained 2,422 lb/yd^3 (1,437 kg/m^3) of cement and 9-percent steel wool and steel fibers combined.

Collepardi et al. reported that the replacement of fine ground quartz sand with an equal volume of well-graded natural aggregate with a maximum size of 0.3 inches (8 mm) did not change the compressive strength at the same water-cement ratio. [35]

Coppola et al. investigated the influence of high-range water-reducing admixture type on the compressive strength. They reported that acrylic polymer admixtures allowed the use of lower water-cement ratios and resulted in higher compressive strengths compared with naphthalene and melamine admixtures. [36]

In a study of the durability of UHPC, Teichmann and Schmidt used the mix proportions shown in table 3. [37] Mix 1 had a maximum aggregate size of 0.32 inches (8 mm) provided by the sand. Mix 2 had a maximum aggregate size of 0.32 inches (8 mm) provided by the basalt.

Table 3. UHPC mix proportions from Teichmann and Schmidt [37]

	Mix 1		**Mix 2**	
Material	**lb/yd^3**	**kg/m^3**	**lb/yd^3**	**kg/m^3**
Cement	1,235	733	978	580
Silica Powder	388	230	298	177
Fine Quartz 1	308	183	503	131
Fine Quartz 2	0	0	848	325
HRWR	55.5	32.9	56.2	33.4
Sand	1,699	1,008	597	354
Basalt	0	0	1,198	711
Steel Fibers	327	194	324	192
Water	271	161	238	141
Water-Binder Ratio	0.19	0.19	0.21	0.21

Researchers at the U.S. Army Corps of Engineers Engineer Research and Development Center have reported on a UHPC-class material referred to as Cor-Tuf. [38,39] The proportions of this UHPC are presented in table 4.

Table 4. UHPC mix proportions of Cor-Tuf by weight [38,39]

Material	Proportions
Portland Cement	1.0
Sand	0.967
Silica Flour	0.277
Silica Fume	0.389
HRWR	0.0171
Steel Fibers	0.310
Water	0.208

Researchers led by Rossi at the Laboratoire Central des Ponts et Chaussees (LCPC) in Paris developed a UHPC-class material referred to as CEMTECmultiscale. [40] The proportions of this UHPC are presented in table 5.

Table 5. UHPC mix proportions for CEMTECmultiscale [40])

Material	lb/yd^3	kg/m^3
Portland Cement	1,770	1,050
Sand	866	514
Silica Fume	451	268
HRWR	74	44
Steel Fibers	1,446	858
Water	303	180

Mixing and Placing

Graybeal has summarized the mixing of UHPC as follows:

> Nearly any conventional concrete mixer will mix UHPC. However, it must be recognized that UHPC requires increased energy input compared to conventional concrete, so mixing time will be increased. This increased energy input, in combination with the reduced or eliminated coarse aggregate and low water content, necessitates the use of modified procedures to ensure that the UHPC does mot overheat during mixing. This concern can be addressed through the use of a high-energy mixer or by lowering the temperatures of the constituents and partially or fully replacing the mix water with ice. These procedures have allowed UHPC to be mixed in conventional pan and drum mixers, including ready-mix trucks. (p. 2) [1]

Mixing times for UHPC range from 7 to 18 minutes, which are much longer than those of conventional concretes. [41,42] This impedes continuous production processes and reduces the capacity of concrete plants. Mixing time can be reduced by optimizing the particle size distribution, replacing cement and quartz flower by silica fume, matching the type of HRWR and cement, and increasing the speed of the mixer. [42] The mixing time can also be reduced by dividing the mixing process into two stages. High-speed mixing for 40 seconds is followed by low-speed mixing for 70 seconds, for a total time of about 2 minutes. [41]

The method of placing UHPC has an influence on the orientation and dispersion of the fibers. [43] The orientation did not affect the first cracking load but had an effect of up to 50 percent on the ultimate tensile strength in bending. The highest strengths were achieved when placement was made in the direction of the measured tensile strength. Stiel et al. reported significant differences between horizontally and vertically cast beams when tested in three-point bending. [44] The fibers in the vertically cast beams were aligned in layers normal to the casting direction. As a result, the splitting and flexural strengths were only 24 and 34 percent of the corresponding values for the horizontally cast beams. However, in a 39-inch (1-m)-thick slab, the fibers were arranged randomly. The orientation of the fibers did not have a significant effect on the compressive strength and modulus of elasticity.

Graybeal has summarized the placement of UHPC as follows:

> The placement of UHPC may immediately follow mixing or be delayed while additional mixes are completed. Although the dwell time prior to the initiation of the cement hydration reactions can be influenced by factors such as temperature and

chemical accelerators, it frequently requires multiple hours before UHPC will begin to set. During extended dwell time, the UHPC should not be allowed to self-desiccate.

Casting of fiber-reinforced concretes requires special considerations in terms of placement operations. UHPCs tend to exhibit rheological behaviors similar to conventional self-consolidating concretes, thus possibly necessitating additional form preparation but also allowing for reduced during-cast efforts. Internal vibration of UHPC is not recommended due to fiber reinforcement, but limited external form vibration can be engaged as a means to facilitate the release of entrapped air. (p. 3) [1]

For the UHPC beams used on the Route 624 bridge over Cat Point Creek in Richmond, VA, the contractor was required to use a plant that was prequalified for UHPC production, and a representative from the UHPC producer was required to be present. [45] The UHPC was mixed in 4-yd^3 (3-m^3) batches in an 8-yd^3 (6-m^3) twin shaft mixer and discharged into a ready-mixed concrete truck for delivery. About 20 to 25 minutes were required to load the mix, mix the UHPC, and discharge the mixer.

During discharge from the truck, cement balls were observed in the mix. This was attributed to exposure of the bags to moisture during storage. The mix was discharged into one end of the beam and allowed to flow. Only limited external vibration was applied for 1 or 2 seconds.

Curing

Curing of UHPC considers two distinct components, specifically temperature and moisture. As with any cementitious composite material, maintaining an appropriate temperature is critical to achieving the desired rate for the cementitious reactions. In addition, given the low water content in UHPC, eliminating loss of internal water by sealing the system or maintaining a high humidity environment is also critical.

The curing of UHPC occurs in two phases. [1,46] Given that UHPC tends to exhibit a dormant period prior to initial setting, the initial curing phase consists of maintaining an appropriate temperature while precluding moisture loss until setting has occurred and rapid mechanical property growth is occurring. The second curing phase may or may not include elevated temperature conditions and a high moisture environment, depending on whether accelerated attainment of particular material characteristics is desired.

Graybeal reported on an extensive program to determine material properties of UHPC using four different post-set curing procedures. [22] These involved steam curing at 194 °F (90 °C) or 140 °F (60 °C) for 48 hours, starting about 24 hours after casting; steam curing at 194 °F (90 °C), starting after 15 days of standard curing; and curing at standard laboratory temperatures until test age.

These three steam-curing methods increased the measured compressive strengths and modulus of elastic, decreased creep, virtually eliminated drying shrinkage, decreased chloride ion penetrability, and increased abrasion resistance. The enhancements achieved by the lower steam temperature and delayed steam curing were slightly less than achieved by steam curing at the higher temperature. The specimens steam cured at 194 °F (90 °C) after 24 hours reached their full compressive strengths within 4 days after casting. Section 3 of this report presents more details of the test results.

More recent work by Graybeal has focused on characterizing the performance of ambient-cured UHPC. [47] This research stems from the recognition that accelerated curing in a steam environment is frequently not practical and also that the ambient-cured properties of UHPC are appropriate for many applications.

Ay compared the compressive strength of 4-inch (100-mm) cubes cured by the following three methods: [48]

- Curing in water until 1 hour before testing.
- Curing in water for 5 days followed by air curing.
- Sealing the cubes in plastic sheeting and then storing them at 68 °F (20 °C) until tested.

The UHPC cubes stored in water followed by air curing had slightly higher compressive strengths than cubes cured by the other two methods.

The compressive strength of UHPC can be increased considerably by using post-set heat curing. [49] Heinz and Ludwig showed that the heat curing at various temperatures between 149 and 356 °F (65 and 180 °C) produced 28-day compressive strengths as high as 41 ksi (280 MPa) compared with strengths of 25 and 27 ksi (178 and 189 MPa) when cured at 68 °F (20 °C). Higher curing temperatures resulted in higher compressive strengths. In addition, the strengths at the end of the curing period at about 48 hours after casting were about the same as the corresponding 28-day strengths. The authors also concluded that curing at 194 °F (90 °C) presented no danger of delayed ettringite formation. [49]

Schachinger et al. observed that initial curing at 68 °F (20 °C) for 5 days, followed by heat curing at 122 to 149 °F (50 to 65 °C), was the most favorable combination to achieve high strengths at ages up to 28 days. [50] Compressive strengths in the range of 36 to 43.5 ksi (250 to 300 MPa) were achieved at ages of 6 to 8 years.

Heinz et al. achieved compressive strengths higher than 29 ksi (200 MPa) at an age of 24 hours after 8 hours storage at 68 °F (20 °C) followed by 8 hours at 194 °F (90 °C) in water. [51] Longer periods of initial storage or heat treatment resulted in higher strengths when ground-granulated blast-furnace slag was included in the UHPC. The authors obtained the highest strengths by including fly ash and autoclaving the UHPC for 8 hours at 300 °F (150 °C).

Massidda et al. showed that autoclaving at a temperature of 356 °F (180 °C) and 145 psi (1 MPa) with saturated steam produced higher compressive strengths and flexural strengths compared with specimens cured at 68 °F (20 °C). [52]

Quality Control Testing

Quality control tests for UHPC in the United States have generally used the same or similar tests as those used for conventional concrete or mortar with or without modifications. Both fresh and hardened concrete properties are measured.

The flow of UHPC is frequently measured using ASTM C1437—Standard Test Method for Flow of Hydraulic Cement Mortar. [1,53] This test method is intended for use with mortars exhibiting plastic to flowable behavior, and thus it is frequently appropriate for fresh UHPC. In this test, both initial flow and dynamic flow are measured. The test is completed immediately after mixing to assess consistency among mixes and appropriateness for casting. [1] On the Route 24 bridge over Cat Point Creek, a minimum dynamic flow of 9 inches (230 mm) was sought for satisfactory workability. [45]

As different versions of UHPC are developed for different applications, alternate workability tests will be needed. For stiffer, non-self-consolidating UHPC, the ASTM C143—Standard Test Method for Slump of Hydraulic-Cement Concrete may be appropriate. [54] Scheffler and Schmidt have reported that development of stiff UHPC formulations for applications such as pavement whitetopping is feasible. [55]

The initial and final setting times of UHPC can be longer than those observed for many conventional cementitious materials. The set times are heavily influenced by the curing temperature. [47] Graybeal measured initial setting times ranging from 70 minutes to 15 hours for different UHPC formulations using the American Association of State Highway and Transportation Officials (AASHTO) T 197 test method for penetration resistance. [22,56,57] The corresponding final setting times ranged from 5 to 20 hours.

Compressive strength testing of UHPC is frequently completed using a modified version of ASTM C39—Standard Test Method for Compressive Strength of Cylindrical Concrete Specimens. [58] The test method is modified to include an increased load rate of 150 psi/second (1 MPa/second) in response to the high compressive strength that UHPC exhibits. [47] Appropriate cylinder end preparation is critical because non-flat or non-parallel end surfaces can cause a reduction in observed compressive strength. [1] End surface preparation for cylinders with early age compressive strengths below 12 ksi can be completed using multiple methods, including capping according to ASTM C617. [1,47,59] Higher strength cylinders should have their ends ground to within 0.5 degrees. [58]

Smaller cylinders have been shown to provide strengths equivalent to traditional size cylinders. Graybeal reported that 3- by 6-inch (76- by 152-mm) cylinders exhibited similar strengths to 4- by 8-inch (102- by 203-mm) cylinders while allowing for the use of a significantly reduced testing machine capacity. [22,60] Use of 2- by 4-inch (51- by 102-mm) cylinders was not recommended because of the increased dispersion present in the results.

Research has demonstrated that the ASTM C109—Standard Test Method for Compressive Strength of Hydraulic Cement Mortars (Using 2-inch (50-mm) Cube Specimens) can also be applied to UHPC. [61] Graybeal reported that 2-inch, 2.8-inch, and 4-inch cubes exhibited compressive strengths up to approximately 7 percent greater than those observed from 3- by 6-inch and 4- by 8-inch (76- by 152-mm and 102- by 203-mm) cylinders. [22,60] Similar findings were reported by Alhborn and Kollmorgen. [62]

On the U.S. Route 6 bridge over Keg Creek in Pottawatomie County, IA, UHPC was used in the longitudinal and transverse joints between the concrete deck panels. [63] The Special Provisions for the project required the contractor to cast twelve 3- by 6-inch (75- by 150-mm) cylinders for verification of concrete compressive strength. [64] Three cylinders were to be tested to verify 10.0 ksi (69 MPa) at 96 hours, three to verify 15.0 ksi (103 MPa) for opening the bridge to traffic, and three at 28 days. The remaining three specimens were treated as reserves. Specimens were required to have their ends ground to 1 degree planeness.

For field-cast UHPC joints, the New York State Department of Transportation (NYSDOT) also requires the casting of twelve 3- by 6-inch (75- by 150-mm) cylinders for testing in sets of three. [65] One set is tested at 4 days, one set at 28 days, one set is to be supplied to the NYSDOT, and one set is treated as reserve.

For qualification testing of the proposed UHPC mix, NYSDOT requires that a minimum of sixty-four 2-inch (50-mm) cubes be cast. Testing ages are 4, 7, 14, and 28 days. Minimum compressive strengths of 14.3 ksi (100 MPa) at 4 days and 21.8 ksi (150 MPa) at 28 days are required.

Frölich and Schmidt investigated the repeatability and reproducibility of tests methods for fresh UHPC. [66] They observed that the values of the measured fresh properties were influenced by the time of measurement, mixing equipment, laboratory conditions, operator, and air-void content. The authors concluded that quality control tests should be made 30 minutes after the start of mixing and that flowable consistency should be measured using the slump flow test.

Summary of Materials and Production

The constituent materials of UHPC generally consist of portland cement, fine sand, ground quartz, HRWR, accelerating admixture, steel fibers, and water. As a class, UHPCs have high cementitious materials contents and very low water-cementitious materials ratios. UHPC can be mixed in conventional mixers but the UHPC mixing time is longer than for conventional concrete. The method of placing UHPC has an influence on the orientation and dispersion of the fibers, which influences the tensile properties of the UHPC. The properties of UHPC are affected by the method, duration, and type of curing. As with conventional concrete, heat curing accelerates the development of strength and related properties. Delaying the application of heat for several days can enhance the measured properties, although it may not be compatible with the rapid production in precasting operations. Smaller size cylinders have been used in quality control for measurement of compressive strengths.

Section 3. Mechanical Properties

This section summarizes information about the various mechanical properties that are relevant to the structural design of UHPC components.

It is important to note that the dispersion and orientation of the fiber reinforcement are critical parameters that influence the mechanical behavior of UHPC. The fiber reinforcement serves to resist tensile stresses in the UHPC component both before and after tensile cracking of the UHPC matrix. Post-cracking mechanical response of UHPC is particularly susceptible to degradation from disadvantageous fiber dispersion and/or orientation. Mixing and placing methods can affect the hardened UHPC mechanical response and thus must be appropriately coordinated to ensure acceptable mechanical performance. [4]

Compressive Strength

Compressive strength is an important property in the design of any concrete structure. It is also the property that is most frequently measured. As discussed in the previous section, cylinder and cube compression test methods used for conventional concrete are appropriate for the determination of UHPC compressive strength. Minor modifications to the test and analysis methods may be required.

Graybeal reported the compressive strengths of nearly 1,000 specimens subjected to the following four different curing conditions: [22]

a) Steam curing at 194 °F (90 °C) and 95-percent relative humidity for 48 hours starting about 24 hours after casting.
b) Steam curing at 140 °F (60 °C) for 48 hours starting about 24 hours after casting.
c) Steam curing at 194 °F (90 °C) for 48 hours starting about 15 days after casting.
d) Curing under laboratory conditions (73 °F (23 °C) and ambient humidity).

Most tests were conducted on 3- by 6-inch (76- by 152-mm) cylinders with the ends ground so that they were parallel within 1 degree. Tests generally used the procedures of ASTM C39, except the loading rate was increased to 150 psi/second (1 MPa/s), and a 6.5-inch (165-mm)- diameter spherical bearing plate was used. [67]

The average measured compressive strengths at 28 days for six cylinders cured using methods a, b, c, and d were 28.0, 24.8, 24.8, and 18.3 ksi (193, 171, 171, and 126 MPa), respectively. Density of the UHPC ranged from 150 to 156 lb/ft^3 (2,400 to 2,500 kg/m^3). Within each curing regime, there was a slight increase in compressive strength as the density increased.

Graybeal also investigated the effect of cylinder and cube size on the measured compressive strength using 2- by 4-inch, 3- by 6-inch, 4- by 8-inch, and 3- by 6.5-inch (51- by 102-mm, 76- by 152-mm, 103- by 203-mm, and 76- by 165-mm) cylinders and 2- and 3.94-inch (51- and 100-mm) cubes. [22,60]

The measured strengths were all within 8 percent of the control 3- by 6-inch (76- by 152-mm) cylinder strength. The cubes had compressive strengths about 5 percent higher than the cylinders.

Similar results were also observed by Orgass and Klug. [68] The smaller cylinders and cubes had a larger standard deviation. [22] Magureanu et al. reported that 3.9-inch (100-mm) cubes had a 20-percent lower measured compressive strength than 2.0-inch (50-mm) cubes. [69]

Graybeal also indicated that loading rates between 35 and 245 psi/seconds (0.24 and 1.7 MPa/seconds) had no noticeable effect on the measured compressive strength, modulus of elasticity, and Poisson's ratio. [22]

Skazlic et al. investigated the effect of cylinder size on the compressive strength of 10 different UHPC mixtures. [70] Cylinder diameters were 2.75, 4, and 6 inches (70, 100, and 150 mm) with a length-to-diameter ratio of 2:1. Assuming a 4- by 8-inch (100- by 200-mm) cylinder as a standard, the authors proposed conversion factors of 1.05 to 1.15 for strengths measured on 2.75- by 5.5-inch (70- by 140-mm) cylinders and 0.85 to 0.95 for strengths measured on 6- by 12-inch (150- by 300-mm) cylinders.

Based on a regression analysis of the data for the particular mix tested, Graybeal determined that the compressive strength gain of UHPC cured under standard laboratory conditions can be represented by the equation in figure 1 for any time after 0.9 days. [22]

$$f'_{ct} = f'_c\left[1\text{-}exp\left(-\left(\frac{t - 0.9}{3}\right)^{0.6}\right)\right]$$

Figure 1. Equation. Compressive strength gain at any age after casting from Graybeal [22].

where:

$f'_c\,t$ = UHPC compressive strength at age t days
f'_c = UHPC compressive strength at 28 days
t = time after casting in days

Graybeal recently completed a follow-on study focused on a readily available UHPC that is formulated for use in field-cast connection applications. [47] A single mix design was cured at 105 °F (41 °C), 73°F (23 °C), and 50 °F (10 °C) to assess the rate of compressive mechanical property development. The time to initiation of compressive mechanical strength gain is provided in figure 2. The relationship between curing temperature and compressive strength is provided in figure 3. The fitting parameters relevant to figure 3 are in table 6.

$$t_{start} = \frac{2.8}{\sqrt{T}}$$

Figure 2. Equation. Relationship between curing temperature and initiation of rapid compressive strength gain from Graybeal [47].

where:
$tstart$= time of initiation of strength gain in days
T = curing temperature in degrees Celsius

$$f'_{c,t} = f'_{c,28d}\left(1 - e^{-\left(\frac{t-t_{start}}{a}\right)^{b}}\right)$$

Figure 3. Equation. Relationship between time after mix initiation and compressive strength as a function of curing temperature from Graybeal [47].

where:

$f'_{c\,28\,d}$ = compressive strength at 28 days
$f'_{c,\,t}$ = compressive strength at time t in days after mix initiation
t_{start} = time of initiation of strength gain in days
a = fitting parameter in days
b = dimensionless fitting parameter

Table 6. Parameters relevant to equation presented in figure 3

Curing Regime	T (°C)	$f'_{c,28d}$ (ksi)	a (days)	b
105°F (41 °C)	41	24.5	0.25	0.25
73° F (23 °C)	23	24	1.0	0.30
50° F (10 °C)	10	22.5	4.0	0.50

Note: 1 ksi = 6.89MPa and °F = 1.8 X °C + 32.

Kazemi and Lubell also investigated compressive strength as a function of time after casting. [27] The response of a locally sourced UHPC from central Canada was found to correspond to the relationship in figure 3, with a equal to 4 and b equal to 0.5 or 0.6 depending on the fiber content.

Schmidt and Fröhlich reported that irregularities in the loaded surface of specimens tested in compression caused a more pronounced decrease in the measured compressive strength in UHPC than was evident with conventional concrete. [71]

Tests of UHPC in axial compression at elevated temperatures showed that the measured compressive strength decreases as the concrete temperature at testing increases. [72,73] However, some or all of the strength is recovered after the specimens cool down.

Richard reported that compressive strengths as high as 80 ksi (550 MPa) can be achieved at atmospheric pressure and heat treating at 480 °F (250 °C). [12] With pressure, compressive strengths as high as 117 ksi (810 MPa) are possible. With conventional production capabilities and curing at 194 °F (90 °C), strengths of 40 ksi (280 MPa) can be achieved.

Tests of UHPC under biaxial compression have been reported by Curbach and Speck and Leutbecher and Fehling. [74,75]

Additional compressive strength data are available in many of the publications about research and applications of UHPC. These data indicate that the initiation of strength gain and subsequent rate of strength gain depend on the particular UHPC constituent materials, mix proportions, and the curing conditions.

Tensile Strength

In conventional structural design for concrete bridges, the tensile strength of concrete is assumed to be zero in reinforced concrete design and often taken as $6\sqrt{f_c'}$ in prestressed concrete girder design. [76]

The tensile strength of UHPC is higher than that of conventional concrete, and UHPC can exhibit sustained tensile strength after first cracking. The results of tests for tensile strength of UHPC, therefore, often report a value of first cracking strength as well as a peak post-cracking strength. Consequently, tensile strength takes on increasing importance as a property to consider in design.

An example tensile stress-strain response obtained from a readily available UHPC containing 2 percent by volume steel fiber reinforcement was captured by Graybeal and is shown in figure 4. [77] The results shown were developed as part of a study. [78,79]

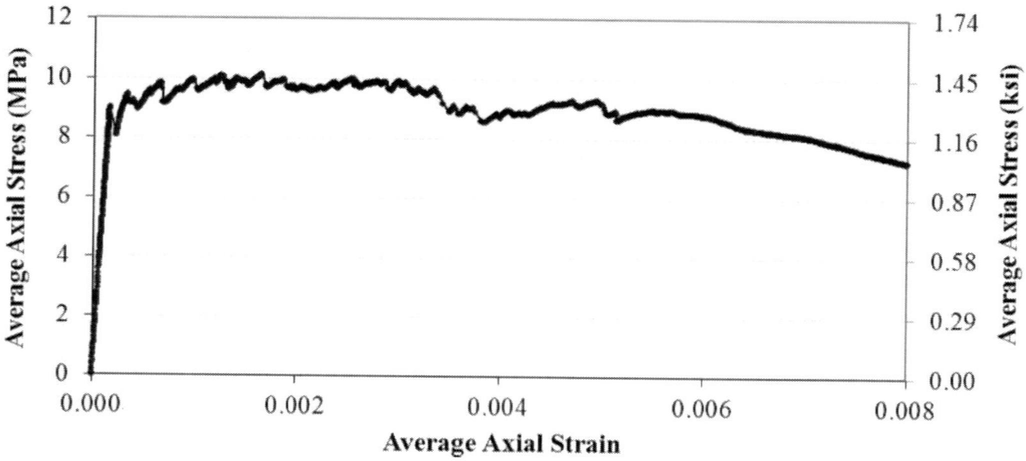

Figure 4. Graph. Tensile stress-strain response of UHPC [77].

Graybeal has proposed the idealized tensile stress-strain response shown in figure 5. [79] This response is based on direct tension tests of two UHPCs with multiple fiber contents. It is proposed as a conceptual illustration of the precracking and postcracking tensile stress-strain response of strain-hardening fiber reinforced concretes, such as UHPC. The behavior is divided into four phases. Phase I is elastic behavior. Phase II is the phase wherein multiple tightly spaced cracks form in the UHPC matrix. The cracks occur individually as the stress in the matrix exceeds the matrix cracking strength. Phase III begins at the strain level where additional cracking between existing cracks is unlikely. Individual cracks widen in this phase. Lastly, Phase IV begins when an individual crack has reached its strain limit and the fibers bridging that crack begin to pull out of the matrix. In a strain-hardening fiber-reinforced concrete, the fiber bridging strength where localization occurs is greater than the cracking strength where multicracking occurs.

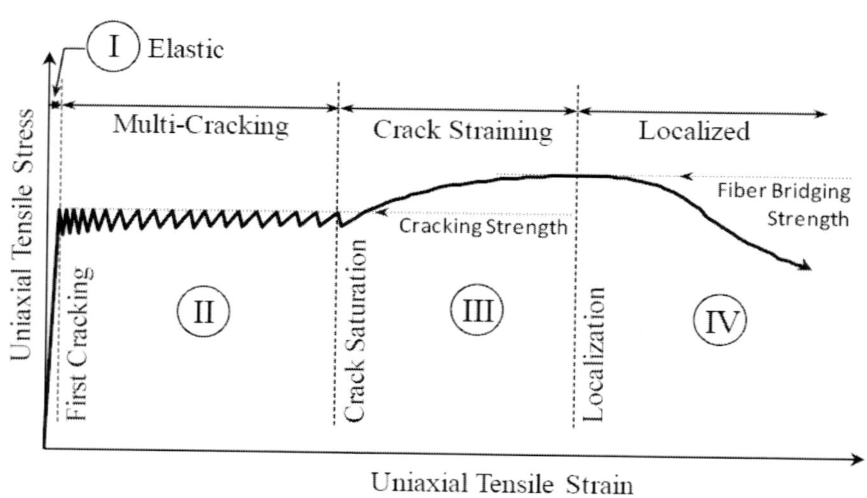

Figure 5. Graph. Idealized uniaxial tensile mechanical response of a UHPC [79].

Standard tensile test methods designed to assess the cracking strength of conventional concrete may be appropriate for assessing the first cracking strength of UHPC, but are unlikely to be appropriate for quantitatively assessing the post-cracking tensile response of UHPC. [22] The ASTM C78—Standard Test Method for Flexural Strength of Concrete (Using Simple Beam with Third-Point Loading) and ASTM C496—Standard Test Method for Splitting Tensile Strength of Cylindrical Concrete Specimens fall into this category. [80,81] Both test methods include assumptions of mechanical behaviors that are not consistent with strain-hardening fiber- reinforced concretes and thus are likely to overestimate the tensile strength of the UHPC. Graybeal has proposed a modified version of ASTM C496. [82] The modified test method includes a requirement to monitor the first cracking of the UHPC during the test and calculate the splitting tensile strength based on the observed first cracking load.

Flexure-based test methods have been proposed, and in some cases, standardized. ASTM C1018 (withdrawn), ASTM C1609, and RILEM TC 162-TDF all present test methods for use in determining the tensile response of fiber-reinforced concretes. [83,84,85] Methods have been proposed for analyzing the test results so as to develop uniaxial tensile response curves. (See references 4, 85, 86, and 87.) However, these types of flexure tests have been demonstrated to be susceptible to overindications of strength as a result of the use of inappropriate support conditions. [88]

A variety of direct tension test methods have been developed. In a direct tension test, the UHPC specimen is loaded in uniaxial tension and thus the tensile response can be directly captured by measuring the load on and the strain experienced by the specimen. Direct tension tests can be divided into two groups, namely, tests that allow rotation of the ends of the test specimen and tests that do not. The tests with rotation might provide an indication of first cracking strength, but are not appropriate for assessment of post-cracking behaviors. This is because local inconsistencies in stiffness at the plane of the first crack result in rotation and fiber pullout at this crack prior to the generation of a full set of additional cracks. Fixed-end tests that do not allow rotation at cracks are appropriate for capturing the full tensile stress-strain response. However, these tests are difficult to complete because of the bending stresses that can be imparted to the specimen during initial setup.

Graybeal reported measurements of tensile strength using flexural prisms, split cylinders, mortar briquettes, and direct tension tests of cylinders. [22] The combined results of these tests indicated a first tensile cracking strength of approximately 1.3 ksi (9.0 MPa) for steam-cured specimens and approximately 0.9 ksi (6.2 MPa) without any heat treatment.

The modulus of rupture values for first cracking determined by the ASTM C1018 prism flexure test varied from 1.3 to 1.5 ksi (9.0 to 10.3 MPa), depending on the method of steam curing, and had an average value of 1.3 ksi (9.0 MPa) for untreated specimens. [83] These specimens exhibited large deflections before the post-cracking peak load was reached.

In the split cylinder tests (ASTM C496), measured splitting tensile strengths at first cracking were 1.7 ksi (11.7 MPa) for steam-cured specimens and 1.3 ksi (9.0 MPa) for untreated specimens. [81] For the steam-cured specimens, the splitting tensile strengths at first cracking varied from 3 to 5 percent of the measured compressive strength. The post-cracking peak tensile splitting stresses ranged from 12 to 16 percent of the compressive strength.

First cracking tensile strengths using briquettes in accordance with AASHTO T 132 ranged from 1.0 to 1.4 ksi (6.9 to 9.7 MPa), depending on the method of steam curing. For untreated specimens, the average value was 0.9 ksi (6.2 MPa). [89]

In the direct tensile tests of 4- by 8-inch (102- by 203-mm) cylinders, first tensile cracking occurred between 1.1 and 1.6 ksi (7.6 and 11.0 MPa), depending on the method of steam curing, and between 0.8 and 1.0 ksi (5.5 and 6.9 MPa) for untreated specimens.

In this study, Graybeal concluded that the tensile strength (*fft*) of UHPC can be related to the measured compressive strength (f'_f) by the equation in figure 6.

$$f_{ct} = 7.8\sqrt{f_c'} \text{ or } 8.3\sqrt{f_c'} \text{ depending on the method of steam curing}$$

$$f_{ct} = 6.7\sqrt{f_c'} \text{ for untreated specimens}$$

Figure 6. Equation. Concrete tensile strength approximations.

Subsequent research by Graybeal and Baby on this topic has resulted in the development of a uniaxial direct tension test method applicable to UHPC. [79] This test method, whose concept is based on a standard tension test applied to metals, provides the uniaxial tensile mechanical response of UHPC and is applicable to both cast and extracted test specimens. Tests were completed on two UHPCs containing multiple steel fiber reinforcement percentages and cured through ambient laboratory and steam-treated conditions. The results demonstrated that these two UHPCs could sustain more than 1.3 ksi (9 MPa) of uniaxial tensile load through a strain of at least 4,000 millionths.

Baby, Graybeal, Marchand, and Toutlemonde investigated the use of flexural tensile test methods for UHPC and the associated analyses necessary for appropriate interpretation of the results. [86] These analyses, often referred to as inverse analyses, derive the uniaxial tensile response from the observed load, deflection, and possibly surface strains observed during a flexure prism test. This research demonstrated that flexure test methods can be applied, but that capture of specific response observations are necessary and appropriate interpretation of data is critical. This research was conducted alongside the research presented in Graybeal and Baby, allowing direct comparison of results. [79]

Reineck and Frettlöhr investigated the effect of specimen size on the flexural and axial tensile strengths. [90] The depth of the flexural specimens ranged from 1 to 6 inches (25 to 150 mm) with width-to-depth ratios ranging from 1 to 5. The same sizes and ratios were used for the axial tension tests. The authors reported a decrease in both strengths with increasing size of the test specimens.

Schmidt and Fröhlich observed that specimens heat cured at 194°F (90 °C) for 48 hours and tested in flexure had a 15-percent higher flexural strength than specimens stored continuously at 60 °F (20 °C). [71]

A study by Wille and Parra-Montesinos investigated the effects of beam size, casting method, and support conditions on UHPC flexure test results. The study reported that large discrepancies in results were possible for an individual UHPC, depending on the test setup and specimen characteristics. [88]

Flexural strengths have also been reported by others. (See references 39, 69, 91, 92, 93, and 94.)

Axial tension tests have also been reported by others. (See for example references 75, 95, 96, 97, 98, 99, 100, and 101.)

Biaxial compression-tension tests have been reported by Leutbecher and Fehling and by D'Alessandro et al. [75,102]

Modulus of Elasticity

Graybeal measured the modulus of elasticity in compression in accordance with ASTM C469 at ages from 1 to 56 days for cylinders cured according to the four regimes described under Compressive Strength. [22,103] Reported values were mostly the average value of six cylinders. After steam curing, the measured values were about 7,250 ksi (50 GPa). Cylinders cured under standard laboratory conditions had modulus of elasticity values of about 6,200 ksi (42.7 GPa) at 28 days. In terms of strength, modulus of elasticity, and strain at peak load, the UHPC showed very little change after completion of steam curing. The specimens cured under laboratory conditions continued to gain strength for at least 8 weeks after casting but the increase in modulus of elasticity and the decrease in strain at peak load seemed to stop at about 1 month.

The modulus of elasticity was also measured in direct tension tests. The average measured values were 7,500 ksi (51.9 GPa) for steam-treated specimens and 6,900 ksi (47.6 GPa) for untreated specimens. These values were slightly higher than measured in compression.

The equation in figure 7 or modulus of elasticity was proposed by Graybeal based on the general form of the AASHTO equation and values of f'_c between 4.0 and 28.0 ksi (28 to 193 GPa). [22]

$$E_c = 46200\sqrt{f'_c} \quad \text{in psi units}$$

Figure 7. Equation. Graybeal equation for UHPC modulus of elasticity [22,103].

where:

E_c = modulus of elasticity

f'_c = UHPC compressive strength

Subsequent research by Graybeal has developed additional results related to the modulus of elasticity of UHPC. [47] These tests, completed on a UHPC specifically formulated for use as a field-cast material in connections between structural components, found that the equation in figure 8 is inappropriate for strengths between 14 and 26 ksi (97 and 179 MPa). This research also investigated the impact of curing temperature on compressive mechanical response development and found that the modulus of elasticity is related to the compressive strength and is largely independent of curing temperature.

$$E_c = 49000\sqrt{f_c'} \text{ in psi units}$$

Figure 8. Equation. Graybeal equation for UHPC modulus of elasticity [47].

M a et al. developed the equation in figure 9 for U H PC containing no coarse aggregates: [104]

$$E_c = 525,000\left(\frac{f_c'}{10}\right)^{\frac{1}{3}}$$

Figure 9. Equation. Ma et al. equation for UHPC modulus of elasticity [104].

Modulus of elasticity values have also been reported by others. (See references 91, 92, 94, and 105.)

Diederichs and M ertzsch measured the stress-strai n rel ati onshi ps at concrete temperatures ranging from 68 to 1,560 °F (20 to 850 °C). [73] They observed a large reduction in both strength and modulus of elasticity at the higher testing temperatures. At the same time, the strain at peak stress increased at the higher temperatures. The same observations were made by Pi mi enta et al . for temperatures ranging from 68 to 1,110 °F (20 to 600 °C). [72] However, some or all of the loss was recovered after the specimens cooled down.

Poisson's Ratio

Table 7 lists values of Poisson's ratio determined by various researchers.

Table 7. Values of Poisson's ratio

Poisson's Ratio	Reference (First Author)
0.2	Simon [106]
0.16	Joh [107]
0.21	Ahlborn [108]
0.19	Bonneau [105]
0.18	Graybeal [109]
0.18	Ozyildirim [45]

Fatigue Behavior

Ocel and Graybeal reported a fatigue test of an AASHTO Type II girder. [110] The upper limit of the fatigue load was just below the static load levels that would cause flexural and shear cracking of the girder. The first cracks were observed after 0.64 million cycles at the intersection of the web and bottom flange in one shear span. After 1.405 million cycles, the same cracks were observed in the other shear span. Flexural cracks were noticed in the constant moment region after 1.888 million cycles. These were accompanied by a longitudinal crack in the bottom flange. Testing continued to 12 million cycles, during which the existing cracks continued to lengthen and additional cracks occurred, but there was no indication of fatigue degradation or change in the global behavior of the girder.

Prior to construction of a UHPC bridge in the city of Calgary, a 39-inch (1-m)-long transverse section was tested in flexural fatigue. [111] The section was subjected to 1 million cycles between 20 and 80 percent of the design service load, 1 million cycles between the 20 and 80 percent of the observed first cracking load, and 1 million cycles between 20 and 80 percent of the failure load for companion sections that contained fiber-reinforced plastic reinforcing bars. Following the fatigue testing, the specimen was loaded to failure with the maximum load being greater than expected.

Prior to construction of a UHPC waffle-slab bridge deck in Wapello County, IA, tests were conducted on specimens representing a full-scale portion of the bridge. [112,113] A single point load representing a wheel load was placed at two critical locations. No fatigue damage was noted after 1 million cycles of loading at each location.

Graybeal and Hartmann conducted flexural fatigue tests on 2-inch (51-mm)-square beams. [114] In one set of tests, uncracked specimens were loaded to produce different stress ranges. Most specimens survived more than 6 million cycles of loading. In a second series of tests, the specimens were precracked and then tested in fatigue with loads cycling from 10 to 60 percent of the cracking load. One specimen failed after 9,950 cycles, while the other failed after 129,700 cycles. In these tests, some of the steel fiber reinforcement was observed to have fractured rather than pulling out of the UHPC matrix.

Schmidt et al. investigated the fatigue behavior of UHPC cylinders loaded in axial compression at various stress range levels. [115] They observed that specimens with a ratio of stress range to compressive strength of 0.45 survived 2 million load cycles without failure. The test specimens that survived the 2 million cycles of loading had only a slight decrease in compressive strength compared with specimens without any preceding load cycles.

Fatigue tests of UHPC specimens under various combinations of stress level and stress range by Fitik et al. showed a range of cycles to failure from 2.5 to more than 7.0 million cycles. [116] They attributed the wide range to local faults, which initiated the failure process.

Uniaxial compression tests reported by Grünberg et al. and Lohaus and Elsmeier with a minimum stress limit of 5 percent of the static strength and varying upper stress level resulted in the number of cycles to failure ranging from about 2.5 to 7.1 million. [117,118]

Behloul et al. conducted flexural fatigue tests on 4- by 4- by 16-inch (100- by 100- by 400-mm) prisms made of two different UHPC formulations. [119] Prior to fatigue loading, the specimens were loaded to produce a crack opening of 0.012 inches (0.3 mm). The specimens were then cycled at 5 Hz between 10 and 90 percent of the first cracking strength. After 1 million cycles, the specimens were loaded statically, and the results were compared with specimens not subjected to fatigue loading. The fatigue loading appeared to have no effect on the overall mechanical behavior.

Lappa et al. reported flexural fatigue tests of 5- by 5- by 40-inch (125- by 125- by 1,000-mm) beams with a maximum load equal to about 75 percent of the static strength. [120] The number of cycles to achieve fatigue fracture ranged from 29,295 to 170,771.

Thermal Properties

Coefficient of Thermal Expansion

Coefficients of thermal expansion (COTE) measured by various researchers are shown in table 8.

Table 8. Values of coefficients of thermal expansion

COTE		Reference First Author
Millionths/°F	Millionths/°C	
8.2 to 8.7	14.7 to 15.6	Graybeal [22]
6.7	12	Fehling [121]
5.6 to 6.7	10 to 12	Simon [106]
7.6 to 8.2	13.6 to 14.8	Ahlborn [108]
6.7	12	Behloul [122]

The French Interim Recommendations state a value of 6.1 x 10^{-6}/°F (11 x 10^{-6}/°C) if no other value can be determined. [4]

Heat of Hydration

Graybeal measured heat of hydration in a well-insulated calorimeter and reported a temperature rise of about 65 °F (36 °C). [22]

Bond Strength

Carbonell et al. investigated the bond strength between conventional concrete substrates and UHPC toppings. [123] Primary variables were surface temperature and moisture condition of the substrate. Half the specimens were subjected to 300 freeze-thaw cycles in accordance with ASTM C666 Method B. [124] The authors evaluated bond strength using an indirect splitting tensile test along the interface. Samples subjected to the freeze-thaw tests had greater bond strength than samples of the same age without freeze-thaw cycles. Samples in which the substrate was saturated before placing the UHPC achieved higher bond strengths than samples with a dry substrate.

Impact Resistance

Bindiganvile et al. compared the impact resistance of UHPC with that of conventional fiber reinforced concrete (FRC). [125] Under quasistatic loading, UHPC was two to three times stronger in flexure and absorbed three times greater energy than conventional steel FRC or polypropylene FRC. Under impact loading, the UHPC was approximately twice as strong as conventional FRC and dissipated three to four times as much energy.

Cadoni et al. observed that the first cracking stress under dynamic loading was two to three times greater than under static load. [126] The impact resistance of UHPC for use in piles was investigated by Leonhardt et al. [127]

Further discussion of impact resistance can be found in section 6 under the discussion of security applications.

Creep

The standard test method for creep in North America is ASTM C512. [128] In this test, specimens are subjected to a constant axial stress, and the change in length over time is measured. Results may be expressed as a creep coefficient = creep strain/initial strain or specific creep = creep strain/applied stress.

Graybeal conducted creep tests on 4-inch (102-mm)-diameter cylinders loaded at ages of 4, 21, and 28 days depending on the method of curing. [22] Creep coefficients after 1 year ranged from 0.29 to 0.78, and specific creep ranged from 0.04 to 0.15 millionth/psi (5.7 to 21.2 millionths/ MPa), depending on the method of curing and loading age. For reference, the specific creep of conventional concrete is in the range of 0.25 to 1.0 millionths/psi.

Graybeal also conducted creep tests on 4-inch (102-mm) diameter cylinders with compressive strengths between 8.0 and 13.0 ksi (55 and 90 MPa) at stress-to-strength ratios ranging from 0.60 to 0.85 to represent the application of prestressing forces prior to steam curing. Measured creep coefficients after 30 min under sustained load ranged from 0.32 to 0.85. These values would be considered high for the short duration of loading.

Burkhart and Müller measured the effect of age of loading, specimen size, stress level, and curing conditions (sealed and unsealed) on the creep of UHPC. [129] Reported specific creep values after about 100 days under load ranged from 0.11 to 025 millionths/psi (16 to 35 millionths/MPa). Measured creep coefficients after 100 days under load were between about 0.9 and 1.3. The measured creep was observed to decrease with age at loading and increased specimen size. This behavior is similar to that of conventional concrete.

Ichinomiya et al. reported specific creep values ranging from 0.19 to 0.28 millionths/psi (28 to 40 millionths/MPa) after 150 days under load for specimens loaded at 2 and 4 days. [92] For specimens loaded at 28 days, the specific creep was about 0.08 millionths/psi (11 millionths/MPa) after about 120 days.

Acker and Behloul reported specific creep values between 0.30 and 0.22 millionths/psi (43 and 32 millionths/MPa) for ages of loading between 4 and 28 days. [130] Fehling et al. reported specific creep values between 0.32 and 0.15 millionths/psi (47 and 22 millionths/MPa) and creep coefficients between 2.27 and 1.08 for ages of loading between 1 and 28 days. [121]

Francisco et al. reported creep strains of about 1,000 millionths after 30 days under load at a stress of about 8.7 ksi (60 MPa), corresponding to a specific creep of about 0.12 millionths/psi (17 millionths/MPa). [131] The 2.75-inch (70-mm)-diameter cylinders were cured at 122 °F (50 °C) prior to loading at an age of 2 days. Drying creep was negligible.

Francisco et al. showed that the specific creep was about the same for heat-treated UHPC specimens loaded at an age of 2 days to 25 and 40 percent of the compressive strength. [132] Flietstra et al. investigated the creep caused by applying a compressive stress and then subjecting the loaded specimens to different curing regimes. [133] This test simulated transfer of the prestressing force prior to heat treatment.

Shrinkage

Two types of shrinkage may be present in UHPC. Drying shrinkage is that caused by loss of moisture from the UHPC. Autogenous shrinkage is that caused by a decrease in volume as the cementitious materials hydrate. The standard test in the United States for measuring shrinkage is ASTM C157, which is designed to measure drying shrinkage beginning after the concrete has hardened. [134] Other methods are used to measure autogenous shrinkage because these measurements must begin immediately after the UHPC is placed.

Shrinkage of UHPC measured in accordance with ASTM C157 using 3- by 3-inch (76- by 76-mm) prisms provided an ultimate shrinkage range of 620 to 766 millionths, depending on the method of steam curing, and 555 millionths for untreated specimens. [22] The initial shrinkage rate of UHPC was also measured in separate tests. During the initial hydration period, peak shrinkage of 64 millionths/hour was measured. As much as 400 millionths of shrinkage occurred in the first 24 hours for untreated specimens. Following steam curing, further shrinkage was almost eliminated. [22,130]

Measurements of shrinkage by Burkhart and Müller starting 1 or 2 days after casting showed no difference between sealed and unsealed cylinders and among specimens with diameters of 3, 4, and 6 inches (75, 100, and 150 mm). [129] They attributed most of the shortening to that caused by autogenous shrinkage, with very little caused by drying shrinkage. All values were about 300 millionths after 200 days of measurements. Autogenous shrinkage values of 600 to 900 millionths at 28 days were reported by Eppers and Müller, 200 to 550 millionths at 150 days by Ichinomiya et al., and about 640 millionths at 365 days by Lallemant-Gamboa et al. [135,92,93]

Fehling et al. reported total shrinkage of 700 and 900 millionths at 7 and 28 days, respectively, for sealed specimens. For specimens subjected to heat treatment, the subsequent shrinkage was negligible. [121]

Francisco et al. reported autogenous shrinkage of about 270 millionths and drying shrinkage of about 100 millionths at 350 days on 2.75-inch (70-mm)-diameter cylinders cured at 122 °F (50 °C). [131]

Ma et al. showed that the autogenous shrinkage could be reduced considerably by including a basalt coarse aggregate with an aggregate size ranging from 0.08 to 0.40 inches (2 to 5 mm). [104] The coarse aggregate had a relatively small effect on the fresh concrete properties, compressive strength, and modulus of elasticity. Significant reduction of early age autogenous shrinkage was obtained by replacing silica fume with metakaolin in specimens cured at 68 °F (20 °C). [136] For UHPC cured at 108 °F (42 °C), the total shrinkage measured for a mix containing metakaolin was negligible compared with mixes with silica fume or fly ash. [110]

To offset the magnitude of autogenous shrinkage, Suzuki et al. and Kim et al. investigated the use of an expansive additive and a shrinkage reducing additive. [137,138] Suzuki et al. reported that an autogenous shrinkage of more than 700 millionths would be reduced to zero with the use of these materials. Kim et al. reported that total shrinkage at 90 days was reduced from 800 to 400 millionths.

Summary of Mechanical Properties

The application of heat curing has a significant and immediate impact on the mechanical properties of UHPC. It increases the compressive strength, tensile cracking strength, and modulus of elasticity. It decreases creep and virtually eliminates subsequent shrinkage. These beneficial properties can also be achieved without heat curing. However, the effect is reduced, and it takes a longer time to achieve the beneficial properties.

Sufficient information has been published about the mechanical properties of UHPC to establish a range of properties to consider in structural design. These are listed in table 9.

Table 9. Range of UHPC material properties

Property	Range	
Compressive strength	20 to 30 ksi	140 to 200 MPa
Tensile cracking strength	0.9 to 1.5 ksi	6 to 10 MPa
Modulus of elasticity	6,000 to 10,000 ksi	40 to 70 GPa
Poisson's ratio	0.2	0.2
	5.5 to 8.5	10 to 15
Coefficient of thermal expansion	millionths/°F	millionths/°C
Creep coefficient1	0.2 to 0.8	0.2 to 0.8
	0.04 to 0.30	6 to 45
Specific creep1	millionths/psi	millionths/MPa
	Up to 900	Up to 900
Total shrinkage2	millionths	millionths

[1] Depends on curing method and age of loading.

[2] Combination of drying shri nkage and autogenous shrinkage and depends on curing method.

Creep of U H PC is much less than conventional concrete. This results in reduced prestress losses but can be detrimental if relied on to reduce stresses in restrained members.

The total shrinkage reported in table 9 includes both drying and autogenous shrinkage. From the reported data, most of the shrinkage is autogenous shrinkage.

U H PC has sufficient fatigue resistance in both tension and compression to resist several million cycles of loading. Its impact strength is two to three ti mes higher than its static strength.

Section 4. Structural Design and Structural Testing

This section summarizes available information about the structural design of UHPC members and the testing that has been performed on structural members. The different sections in this section correspond to articles in the AASHTO Load & Resistance Factor (LRFD) Bridge Design Specifications. [76]

As noted in section 3, it is important to recognize that the dispersion and orientation of the fiber reinforcement are critical parameters that influence the structural behavior of UHPC. The fiber reinforcement serves to resist tensile stresses in the UHPC component both before and after tensile cracking of the UHPC matrix. Post-cracking structural response of UHPC is particularly susceptible to degradation from disadvantageous fiber dispersion and/or orientation. Mixing and placing methods can affect the hardened UHPC mechanical response and thus must be appropriately coordinated to ensure acceptable structural performance. A framework for addressing the reliance on fiber reinforcement in the tensile mechanical resistance of UHPC structural components has been presented in the Service d'étude des transports, des routes et de leurs aménagement-Association Francaise de Genie Civil (SETRA-AFGC) design recommendations. [4]

Flexural and Axial Loads

Flexural Members

The calculated flexural resistance of concrete components is generally based on the conditions of equilibrium of forces and strain compatibility. The usable compressive strain in unconfined concrete is limited to a maximum value of 0.003. The shape of the stress-strain curve may be any shape that results in a prediction of strength in substantial agreement with test results. For simplification, a rectangular stress block for the compression zone is usually assumed. The tensile strength of the concrete is neglected. The applicability of this approach for use with UHPC has been addressed in several articles.

Graybeal tested a 36-inch (0.91-m)-deep AASHTO Type II girder made of UHPC in flexure using four-point bending on a span length of 78.5 ft (23.9 m). [109] (See figure 10.) The girder contained twenty-four 0.5-inch (12.7-mm)-diameter strands. Prior to reaching peak load, the girder achieved a deflection of almost 19 inches (480 mm) and failed by a combination of tensile fracture of the strands and pullout of the fibers.

The peak applied load on the girder produced a bending moment of 38,700 kip-inches (4,370 kN-m). A flexural analysis, assuming a rectangular stress block and that the UHPC carried no tensile forces after cracking, produced a calculated moment capacity of 27,840 kip-inches (3,150 kN-m)—considerably less than the measured strength. Based on analysis of the measured data, Graybeal proposed that flexural capacity could be calculated more accurately assuming the following UHPC stress-strain curve:

- In compression, a linear relationship up to 0.85 times the compressive strength.
- In tension, a rigid-plastic relationship with a conservative value of post-cracking tensile strength and a limiting tensile strain.

For his test data, stress-strain relationships would be as follows:

- In compression, a linear relationship from the origin to a compressive stress of 24 ksi (165 MPa). Based on the graphical depiction, the modulus of elasticity appears to be about 8,000 ksi (55 MPa).
- In tension, a constant stress of 1.5 ksi (10.3 MPa) over a strain range from zero to 0.007.

The flexural strength of the section can then be calculated using a traditional mechanics of materials approach.

Figure 10. Photo. Flexural test of an AASHTO Type II girder made of UHPC.

Graybeal also tested two pi-girders made of UHPC in flexure using four-point bending. [139] The span lengths of the girders were 69 and 45 ft (21 and 13.72 m). The girders achieved deflections of about 10 and 5 inches (250 and 125 mm) before reaching a peak load. The failure mechanisms of the two girders were similar. The fibers began to pull out from the matrix at a crack near midspan. This shed the tensile force from the fibers to the prestressing strands, which then fractured. The moments at first flexural cracking of the girders were 20,600 and 19,500 kip-inches (2,330 and 2,200 kN-m). The ultimate flexural capacities of the girders were 37,600 and 38,190 kip-inches (4,250 and 4,310 kN-m). A third girder that was intended to fail in shear failed in flexure at a moment of 36,720 kip-inches (4,150 kN-m). Based on an analysis in accordance with the AASHTO LRFD Bridge Design Specifications, a 70-ft (21.3-m) girder has a Service III moment demand of 19,940 kip-inches (2,250 kN-m) and Strength I moment demand of 39,600 kip-inches (4,470 kN-m). The cracking moments observed in the experiment were approximately equal to the Service III moment. The measured flexural capacities were, on average, about 5 percent less than the required values, indicating the need for additional flexural reinforcement.

Meade and Graybeal reported the results of sixteen 6-inch (152-mm)-wide, 15-inch (381-mm)- deep rectangular UHPC beams tested in four-point bending over a span length of 16 ft (4.88 m). [140] The test variables were fiber content (0, 1, and 2 percent by volume) and quantity of conventional nonprestressed reinforcement (0.00 to 1.00 percent by area). Measured compressive strengths of the UHPC ranged from 24.7 to 29.4 ksi (170 to 203 MPa).

Beams containing 1- and 2-percent fiber reinforcement had higher first cracking strengths, better post-cracking flexural response, and higher peak loads than beams without fibers. Increasing the fiber content from 1 to 2 percent resulted in stiffer post-cracking response and higher peak loads. The beams containing no fibers failed when flexure-shear cracks extended into the compression region under the load points, leading to a shear failure of the flexural compression block in the shear span. The beams containing the fibers failed when the fibers pulled out across a critical crack and the reinforcing bars ruptured. No concrete crushing was noted.

Visage et al. reported the results of ten 6-inch (152-mm)-square beams tested in flexure. [141] Test variables included compressive strength, amount of flexural reinforcement, volume of steel fibers, and beam length. Test results were compared with traditional methods of estimating moment-curvature relationships. Other flexural tests have been reported by Gröger et al., Frettlöhr et al., and Stürwald and Fehling. [142,143,144]

Adeline and Behloul reported flexural tests of two 49.2-ft (15-m)-long UHPC beams containing only flexural reinforcement. [145] The beams contained eight or four 0.6-inch (15.2-mm)-diameter strands. The beam with eight strands failed by crushing of the UHPC, whereas the beam with two strands failed by strand rupture. Both beams exhibited large deflections before failure. The authors used a nonlinear multilayer program to predict the moment-deflection curves. They obtained very good agreement between the measured and calculated curves in both the elastic and plastic parts of the curves.

Maguire et al. tested two full-size double-tee beams in flexure. [146] The beams contained 0.7-inch (17.8 mm)-diameter strands and UHPC without steel fibers. The measured strengths exceeded the calculated strengths using measured properties and a strength design approach. The authors concluded that the flexural design procedures of the AASHTO LRFD Specifications for I-girders are applicable to UHPC girders.

Steinberg and Reeves examined the reliability of the flexural strength of UHPC AASHTO standard box beams based on the AASHTO LRFD Bridge Design Specifications. [147,148,149] The reliability analysis consisted of a Monte Carlo simulation and the use of the moment-curvature approach to calculate flexural strength. The authors concluded that the use of the AASHTO LRFD Specifications produces a conservative reliability index when applied to UHPC members. For lightly reinforced members, the design may be overly conservative. To rectify this, they suggested using a more advanced analysis method, such as moment-curvature, or increasing the strength reduction factor.

Prior to construction of Malaysia's first UHPC motorway bridge, a prototype I-girder was tested in flexure using a single point load at midspan. [150] The load-deflection curve predicted by finite element modeling matched the measured values very closely.

Sujivorakul developed a flexural model to predict the moment-curvature relationship for doubly reinforced UHPC beams. [151] The model is based on strain compatibility, equilibrium of forces, and the stress-strain relationship of UHPC in tension and compression.

For strength design, Stürwald and Fehling developed a simplified approach. [144] They used a triangular stress block in compression and a rectangular stress block for tension in the UHPC. This approach gave calculated strengths within 5 percent of the measured strength of three beams.

Moment Redistribution

Walsh and Steinberg examined the moment redistribution capacity of four small-scale continuous two-span UHPC beams with no conventional reinforcement. [152] The test results suggested that the moment redistribution of UHPC is comparable to the 20-percent maximum given in the AASHTO LRFD Bridge Design Specifications. [76]

Compression Members

Tue et al. examined the capacity of stub columns of UHPC confined by a steel tube. [153] The load was applied either to the combined steel and UHPC section or to the UHPC section alone. The authors observed that shrinkage of the UHPC produced a gap between the UHPC and the inside of the steel tube. This gap only closed after the stresses exceeded the service level and lateral strains increased considerably. As such, the confinement effect was not as effective as that achieved with conventional concrete.

Empelmann et al. tested six short columns in concentric compression. The 7.9- by 7.9- by 23.6-inch (200- by 200- by 600-mm) columns contained different amounts of longitudinal and transverse reinforcement. [154]

Yan and Feng also tested short UHPC columns with a diameter of 4.3 inches (110 mm) inside steel tubes with a wall thickness of 0.19 to 0.26 inches (5 to 6.5 mm). [155] Measured compressive strengths were greater than calculated by the equation in figure 11.

$$N = f_y A_S + f_c' A_C$$

Figure 11. Equation. Strength of columns.

where:

f_y = tensile yield strength of the steel tube
A_s = cross sectional area of the steel tube
f'_c = compressive strength of 4-inch (100-mm) UHPC cubes
A_c = cross sectional area of the concrete

The authors noted that confinement of the steel tube was not as effective as that for conventional concrete, and therefore, the effect can be neglected in the calculation of axial load.

Tension Members

A method for measuring the uniaxial tensile stress-strain response has recently been developed by a U.S.–French joint project. [78] The test method provides the response for both precracking and post-cracking phases without requiring any complex stress or strain transformations.

Jungwirth and Muttoni reported the results of direct tension tests on dog-bone shaped specimens having a test cross section of 1.8 by 6.3 inches (45 by 160 mm). [156] They reported a linear stress-strain relationship with a modulus of elasticity of about 8,700 ksi (60 GPa) up to a tensile stress of 1.2 ksi (8.5 MPa). Following cracking of the specimens, the tensile stress increased to about 1.45 ksi (10 MPa) before the fibers progressively pulled out at a strain of about 2.5 percent. The authors also performed tests on specimens containing nonprestressed reinforcement ranging from 1 to 4.5 percent by area. All specimens exhibited well-distributed cracking and strains as high as 10 percent.

Further discussion of the tensile response of UHPC is provided in section 3 under the heading of tensile strength.

Bearing

Holschemacher et al. investigated the bearing strength of two UHPCs, with and without helical reinforcement, using two different specimen heights and different diameters of loading area. [157,158] Results were compared with the German Standard DIN 1045-1. [159] The results indicated that the strengths calculated using DIN 1045-01 equations need to be modified by a factor such as 0.8 for them to be applicable to UHPC.

Hegger et al. tested various details for joints between precast UHPC columns. [160] The major test variables included dry and wet joints. With dry joints, the surface treatment of the interface surfaces and the longitudinal reinforcement ratio were variables. With wet joints, the mortar thickness and transverse reinforcement ratios with welded wire reinforcement and steel plates were variables. Measured bearing capacities were slightly less than measured on a continuous reference column.

Shear and Torsion

Sectional Design

The AASHTO LRFD shear design sectional model involves the calculation of three components that contribute to shear resistance. They are the concrete contribution, the transverse or shear reinforcement contribution, and any vertical component of prestressing force from draped strands. The procedure involves a combination of theory and empirical factors. In UHPC beams with no conventional transverse reinforcement, there is no reinforcement contribution. The tensile stresses that develop are carried by the UHPC matrix and steel fibers.

Graybeal tested three 36-inch (0.91-m)-deep AASHTO Type II prestressed UHPC girders in shear. The girders contained no nonprestressed shear reinforcement. [109] Each girder failed in a different manner. The first girder failed owing to a preexisting horizontal crack at the base of the web from a prior flexural test. The second girder failed owing to diagonal tension in the shear region. The third girder failed owing to a combination of diagonal tension and strand slip. Because the girders did not contain any nonprestressed shear reinforcement and no draped strands, Graybeal proposed that the shear capacity could be determined by assuming that all the shear forces are carried by diagonal tension and compression in the web of the girder. The limiting value is the post-cracking tensile strength of the UHPC. A conservative estimate of this value would be required. In addition, it is necessary to determine the state of stress in the girder under dead load and prestressing forces.

Graybeal also tested three 33-inch (838-mm)-deep pi-girders with shear spans of 7.0, 6.0, and 6.0 ft (2.13, 1.83, and 1.83 m) under three-point bending. [113] First, shear cracks appeared at shear loads of 175, 180, and 205 kips (780, 800, and 910 kN). The shear loads at failure were 430, 366, and 510 kips (1,910, 1,630, and 2270 kN). However, the third girder failed in flexure rather than loss of diagonal tensile capacity in the web as occurred in the first two girders. Based on an analysis in accordance with the AASHTO LRFD Bridge Design Specifications, a 70-ft (21.3-m)-span girder has a Service III shear demand of 103.2 kips (459 kN) and a Strength I shear demand of 206 kips (916 kN). [76] The measured shear strengths were at least 75 percent greater than the Strength I demand. Based on the assumption that the girder webs carried all the shear force and that the diagonal tensile force acted uniformly over the relevant cross-sectional area of the webs, the calculated diagonal tensile capacities corresponding to the shear loads at failure were 2.5, 2.1, and 2.9 ksi (17.2, 14.6, and 20.3 MPa) for the three girders.

Maguire et al. reported shear tests of two full-size double-tee beams. [146] The beams contained vertical shear reinforcement consisting of welded wire reinforcement with cross wires for anchorage. The UHPC did not contain any steel fibers. Both girders had a measured shear strength that exceeded the calculated shear strength based on measured material properties. The authors concluded that the AASHTO LRFD Bridge Design Specifications for shear design of I-girders is applicable to UHPC girders. [76]

Baby et al. reported on a study investigating the shear performance of UHPC beams. [163,164] Study variables included prestressed versus nonprestressed beams, the inclusion of stirrups for shear reinforcement, and the inclusion of fiber reinforcement. Supplemental beams were cast and then deconstructed to extract prismatic specimens from the web region for three-point bending tests. These small-scale tests provided an indication of the fiber reinforcement orientation and effectiveness as shear reinforcement in the web. This research found that the shear design recommendations contained within the SETRA-AFGC UHPFRC Design Guidelines were conservative for these beams. [4]

Shear tests of UHPC beams without conventional shear reinforcement were conducted by Bunje and Fehling. [165] All specimens failed in flexure. Other shear tests were conducted by Hegger et al., Hegger and Bertram, Cauberg et al., Fehling and Thiemicke, and Bertram and Hegger. (See references 166, 167, 168, 169, and 170.)

Hegger and Bertram tested 15.7-inch (400-mm)-deep prestressed concrete I-beams with a length of 185 inches (4.70 ^{m}). [167,171] Four series of beams were tested as follows:

- Beams without openings (11 tests).
- Beams with a single web opening (9 tests).
- Beams with several web openings (7 tests).
- Beams with additional shear reinforcement near the openings (9 tests).

The beams in the first three series did not contain any conventional transverse reinforcement.

For beams without openings, the shear strength increased as the fiber content increased. An increase in the prestressing force resulted in an increase in the shear strength. The provision of a single opening reduced the shear strength. However, the strengths of beams with two openings were similar to those of beams with a single opening.

Wu and Han reported tests of 11 reinforced concrete I-girders of which 8 failed in shear. [172] The main variables were fiber volume content, flexural reinforcing steel ratio, section type, and span/ depth ratio. No shear reinforcement was provided in the webs. Based on the test results, a formula for the first diagonal cracking load was developed. The authors concluded that the conventional equations for calculating shear strength are not appropriate and developed an analytical model.

Prior to construction of Malaysia's first UHPC motorway bridge, a prototype I-girder was tested in shear using a single point load at midspan. [150] No conventional shear reinforcement was included. The shear strength predicted by finite element modeling was 17 percent lower than the measured strength.

Punching Shear

Section 9 of the AASHTO LRFD Bridge Design Specifications requires a minimum deck thickness of 7.0 inches (175 mm) unless approved otherwise by the owner. [76] This generally precludes the likelihood of a punching shear failure in a bridge deck. The use of thinner sections with UHPC increases the likelihood of a punching shear failure and, therefore, the need to consider it in design.

Harris and Roberts-Wollmann tested twelve 45-inch (1,140-mm)-square UHPC slabs in punching shear. [173,174] The variables in the program were slab thicknesses of 2.0, 2.5, and 3.0 inches (51, 64, and 76 mm) and loading plate dimensions from 1.0 to 3.0 inches (25 to 76 mm) square. No conventional reinforcement was included. The measured compressive strength of the UHPC was 32.1 ksi (221 MPa). Seven of the specimens failed in punching shear and five in flexure. The authors concluded that the American Concrete Institute (ACI) 318 equation for punching shear predicted the failure loads reasonably well but a modified version of the ACI model for breakout loads provided the best prediction. [162] They also concluded that a 1.0-inch (25-mm) slab thickness should provide sufficient thickness to resist punching shear in bridge deck applications.

Three larger slabs with dimensions of 7.0 by 12.0 ft and 3 inches thick (2.1 by 3.7 m by 76 mm) were loaded with a wheel patch load. These tests represented the top flange of a double-tee section. The slabs all failed in tension.

Toutlemonde et al. investigated the local bending and punching shear performance of two-way ribbed bridge deck elements. [175] Developed as a potential alternate for orthotropic bridge decks, these 15-inch (0.38-m)-deep deck elements were composed of a 2-inch (0.05-m)-thick plate and 13-inch (0.33-m)-tall bi-directional ribs with a 24-inch (0.6-m) center-to-center spacing. This study tested two different commercially available UHPC products. The punching shear capacity of the deck plate was observed to be greater than 157 kips (700 kN) under all conventional loading scenarios. When the wheel patch was reduced to a 7.5 x 10.2 inch (0.19 x 0.26 m) size, the authors observed the punching shear resistance to be between 79 and 94 kips (350 and 420 kN).

Naaman et al. evaluated the effect of fibers on the punching shear response of 7-inch (175-mm)- thick concrete bridge decks with and without reinforcing bars. [176] Three different types of fibers were included. Test results showed that the punching shear resistance, the energy-absorption capacity, and the resistance to spalling of slabs having only two bottom layers of reinforcing bars were significantly better than for the control specimen with four layers of reinforcing bars and conventional concrete. The authors concluded that punching shear resistance can be safely taken as twice that calculated using the procedures of ACI 318-05. [177]

Saleem et al. tested eight single-tee, simple span beams with a depth of 5 inches (125 mm), a top flange width of 12 inches (300 mm), and a span length of 48 inches (1219 mm). [178] A center point load was applied over an area 19.7 inches (500 mm) long by 9.8 inches (250 mm) wide, representing an AASHTO HS20 truck dual tire wheel. All beams contained longitudinal flexural reinforcement but only two beams had shear reinforcement. The dominant mode of failure in the beams was shear.

Aaleti et al. reported punching shear tests on the 8-inch (200-mm)-deep waffle slab system proposed for use on a bridge deck in Iowa. [179] They concluded that the system would not experience punching shear failure under the traditional 10 by 20 inch (254 by 508 mm) wheel loads. The measured punching shear strength was nearly 2.3 times the estimated value using the ACI equation recommended by Harris and Roberts-Wollmann. [174]

Punching shear tests were reported by Joh et al. [107] Tests were made on 63-inch (1,600-mm) square slabs with thicknesses of 1.6 and 2.8 inches (40 and 70 mm) and loaded through plates with dimensions of 2.0 by 2.0, 3.0, 3.9, or 4.9 inches (50 by 50, 75, 100, or 125 mm). The 1.6-inch (40-mm)-thick slabs reached their flexural strength before punching occurred. The 2.8-inch (70-mm)-thick slabs failed by typical punching at the center of the slab. The authors confirmed that the ACI 318 equation for punching shear gave a reasonable estimate of the strength.

Bunje and Fehling conducted punching shear tests of UHPC slabs with thicknesses of 1.2, 1.6, 2.0, and 3.1 inches (30, 40, 50, and 80 mm). [165] The slabs did not appear to contain any conventional flexural reinforcement. All slabs failed in a ductile flexural mode with no punching failure.

Moreillon et al. reported punching shear tests in which the primary variables were slab thickness, reinforcement ratio, and fiber volume. [180] The authors developed a model for predicting the punching shear strength.

Interface Shear

Twenty-four push-off tests were conducted by Banta to determine whether the horizontal shear design equations of the AASHTO LRFD Bridge Design Specifications accurately predict the horizontal shear strength between UHPC and lightweight concrete. [181] The test variables were interface surface characteristics, interface area, and area of reinforcement crossing the interface. The author compared the test results of 19 specimens with a smooth interface with the equations in the 2004 version of the Specifications, assuming a resistance factor of 1.0 and a friction factor of 1.0. [182] Calculated strengths were always greater than measured strengths. It should be noted that the cohesion and friction factors have been revised since publication of the 2004 version of the LRFD Specifications.

Maguire et al. cautioned that the contribution of the contact surface between precast UHPC girders and a cast-in-place conventional concrete deck should be ignored because of the difficulty of roughening the top surface of the UHPC girders. [146]

Crane and Kahn investigated the interface shear capacity of five reinforced tee beams with UHPC for the web and high-performance concrete (HPC) for the top flange. [183] Test variables included interface roughness and amount of interface shear reinforcement. Test results were compared with the shear friction equations of the AASHTO LRFD Bridge Design Specifications. The equations were unconservative in predicting the shear strength of smooth interfaces even with relatively high amounts of shear reinforcement. Consequently, it was recommended that a fluted interface be used.

Hegger et al. conducted direct shear tests on joints between precast elements subjected to various levels of compression. [160] They included dry and wet joints with various types of contact surfaces.

Shear Connections

Graybeal evaluated the use of UHPC in shear connectors between precast deck panels and concrete or steel beams. [184] He tested two full-size beam specimens. The first specimen included frequently implemented details used to connect precast concrete slabs to beams.

Conventional grout was used. The second specimen used simplified connection details in combination with UHPC. The tested UHPC connections eliminated all interference points between the girder and deck connectors by engaging the mechanical strength of the UHPC to carry the loads between the connectors across an otherwise unreinforced plane. Each specimen was subjected to more than 11 million cycles of loading followed by a static test to failure. The applied loads surpassed the design loads required by the AASHTO LRFD Bridge Design Specifications. [76] The author observed no damage in the UHPC connections after they were subjected to 168 psi (1.16 MPa) of cyclic horizontal shear stress and 789 psi (5.44 MPa) of static horizontal shear stress along the minimum shear plane.

Hegger et al. have tested headed stud and continuous shear connectors using push-off tests and a beam test. [185,186,187] The test parameters for the continuous connector push-off tests were steel fiber content, transverse reinforcement ratio, and thickness of the connector.

The amount of steel fibers had a minor effect on the connector strength if a minimum fiber ratio was maintained. The arrangement of transverse reinforcement influenced the connector strength, whereas the thickness of the connector influenced strength and the mode of failure. In the beam test, the plastic moment was developed with no cracks developing at the connector.

Jungwirth et al. and Kohlmeyer et al. also conducted push-off tests of continuous shear connectors. [188,189]

Torsion

Fehling and Ismail tested 7-inch (180-mm)-square beams in pure torsion. [190] The parameters included steel fiber type, steel fiber volume, longitudinal reinforcement ratio, and web reinforcement ratio. The use of longitudinal and transverse reinforcement in combination with the steel fibers provided the biggest increase in ultimate torsion capacity and ductility.

Joh tested three 12-inch (300-mm)-square beams in pure torsion. [191] One beam contained no conventional reinforcement, one beam contained longitudinal reinforcement in the corners, and the third beam contained both longitudinal and transverse reinforcement. The cracking torque and torsional strength were reasonably predicted using thin-walled tube theory modified to account for the tensile strength of the UHPC.

Empelmann and Oettel conducted tests on seven 20-inch (500-mm)-square hollow boxes with a wall thickness of 2 inches (50 mm) at midlength. [192] Test variables included fiber content, longitudinal reinforcement ratio, and transverse reinforcement ratio. Four specimens were loaded in pure torsion. Three specimens were loaded with a combination of torsion and axial force. The experimental results were compared with design equations for conventional concrete members based on a space truss model.

Prestressing

Stress Limits

No recommendations about stress limits to be used in UHPC prestressed concrete members were identified. However, Graybeal reported high creep on cylinders loaded to between 60 and 92 percent of the compressive strength at compressive strength levels between 8.5 and 12.5 ksi (59 and 86 MPa). [22]

Loss of Prestress

Loss of prestressing force includes an instantaneous loss when the strands are released and a time-dependent loss caused by creep and shrinkage of the concrete and relaxation of the prestressing strands. A reasonable estimate of the instantaneous loss can be made if the modulus of elasticity of the UHPC is known accurately. The AASHTO LRFD specifications provides two methods for predicting time-dependent losses: [76]

- Approximate estimate of time-dependent losses.
- Refined estimate of time-dependent losses.

Both estimates rely heavily on empirical methods. The applicability of these methods for use with UHPC needs to be verified because this study identified no direct methods to measure prestress losses in UHPC.

Calculated prestress losses for Type II AASHTO girder based on material property tests were 35.6 ksi (245 MPa). [109] This included 15.4 ksi (106 MPa) for instantaneous loss, 10.0 ksi (69 MPa) for shrinkage, 6.9 ksi (48 MPa) for creep, and 3.1 ksi (21 MPa) for relaxation.

Reinforcement Details

Article 5.10 of the AASHTO LRFD Bridge Design Specifications addresses reinforcement details. [76] No specific publications addressing these details for use with UHPC were identified. It is likely, however, that most of these provisions could be used with UHPC because of UHPC's higher compressive and tensile strengths.

Development and Splices of Reinforcement

Deformed Bars in Tension

New York State Department of Transportation performed pullout tests of No. 4, 5, and 6 bars embedded 2.9, 3.9, and 4.9 inches (75, 100, and 125 mm), respectively, in 15.7-inch (400-mm)- diameter UHPC cylinders, which resulted in reinforcement fracture within the length of bar not cast into the UHPC. [193]

Graybeal and Swenty conducted pullout tests on No. 4 reinforcing bars embedded into 6-inch (152-mm) cubes of two different UHPCs. [194] The rebar was bonded to the field-cast UHPC for 3 inches only, with the remainder of the length debonded by a foam bond-breaker. All of the specimens were cast and cured in ambient laboratory conditions. Pullout tests on a UHPC formulation intended for use in precast concrete applications resulted in pullout of the bar after the tensile yield strength of the bar had been surpassed. Pullout tests on a UHPC formulation intended for field-cast applications resulted in tensile rupture of the reinforcement.

Pullout tests were also performed by Holschemacher et al. using 0.32- and 0.39-inch (8- and 10-mm)-diameter bars. [195,196] They observed that the bond strength and stiffness increased with testing ages. Fehling et al. also performed pullout tests on 0.47-inch (12-mm)-diameter bars with various amounts of concrete cover and embedment lengths. [197]

Hossain et al. completed pullout and development length tests of glass fiber reinforced polymer (GFRP) rebar embedded in two different UHPC formulations. [198] Both No. 5 and 6 bars were tested with both high and low modulus of elasticity GFRP formulations. Larger bars and longer bond lengths were observed to result in lesser bond strengths, with all specimens failing via bar pullout.

Deformed Bars in Compression

No publications about the development length of deformed bars in compression in UHPC were identified.

Lap Splices

Graybeal evaluated the performance of six connection details for use between precast concrete elements. [193,199] Four connections represented transverse joints between full-depth precast concrete deck panels. Two connections represented longitudinal joints between adjacent deck bulb-tee girders. Table 10 provides the reinforcement details used in the connection regions. Bars from adjacent panels were offset by half the bar spacing.

Table 10. Reinforcement used in connections

Orientation	Bar Size	Bar Type	Lap Length, inches	Bar Spacing, inches	
				Top	Bottom
Transverse	No. 5	Headed uncoated	3.5	17.7	7.1
Transverse	No. 4	Hairpin epoxy coated	3.9	4.3	4.3
Transverse	No. 5	Straight galvanized	5.9	17.7	7.1
Transverse	No. 5	Straight uncoated	5.9	17.7	7.1
Longitudinal	No. 5	Headed uncoated	3.5	17.7	7.1
Longitudinal	No. 5	Straight uncoated	5.9	17.7	7.1

The specimens were loaded on a simple span, with the load applied through a simulated wheel patch placed adjacent to the connection near mi dspan. Cyclic loads were applied first, with the test program including at least 2 million cycles to a load just below the cracking strength of the specimen followed by at least 5 million cycles to a load larger than the cracking strength of the specimen. After the completion of the cyclic testing, each test specimen was statically loaded to failure. All the specimens survived 7 million cycles of fatigue loading.

The tests showed that noncontact, lap-spliced reinforcement in the transverse and longitudinal connections was not susceptible to debonding under cyclic and static loads. The development length of straight, uncoated No. 5 reinforcing bars in this test program was demonstrated to be equal to or less than 5.9 inches (150 mm) in a non-contact lap splice configuration.

Hegger et al. reported on direct tension tests of lap-spliced specimens. [160] The test variables were bar diameter, lap length, steel fiber ratio, transverse reinforcement ratio, and concrete cover.

Hossain et al. reported on testing lap-spliced GFRP rebar in field-cast connections between prefabricated bridge deck elements. [198] This testing, which included both static and cyclic flexural loading of the beam splice connections, demonstrated that 5.9- to 8.9-inch (150- to 225-mm) lap splice lengths can be appropriate for GFRP rebar embedded in UHPC.

Standard Hooks in Tension

No publications about the development length of standard hooks in tension in UHPC were identified. However, it is likely that the existing provisions of the AASHTO LRFD Bridge Design Specifications are applicable because of UHPC's higher compressive and tensile strengths. [76]

Welded Wire Reinforcement

No publications about the development length of welded wire reinforcement in UHPC were identified. However, it is likely that the existing provisions of the AASHTO LRFD Bridge Design Specifications are applicable because of UHPC's higher compressive and tensile strengths. [76]

Shear Reinforcement

No publications about the development length of shear reinforcement in UHPC were identified. However, it is likely that the existing provisions of the AASHTO LRFD Bridge Design Specifications are applicable because of UHPC's higher compressive and tensile strengths. [76]

Development of Prestressing Strand

Measured transfer and development lengths from various researchers are summarized in table 11.

Table 11. Measured transfer and development lengths

Strand Diameter		Transfer Length		Development Length		Source
inches	mm	inches	mm	inches	mm	
0.6	15.2	14 8.7 to	356	< 35	< 890	Ruiz et al. [200,201]
0.5	12.7	11.0	220 to 280	—	—	Bertram and Hegger [202]
0.7	17.8	17 to 21	430 to 530	—	—	Maguire et al. [146]
0.5	12.7	—	—	< 37	< 940	Graybeal [109]

— No data reported.

Graybeal reported results of a study investigating the lap-splice length of unstressed prestressing strands. [203] Strands were lapped inside UHPC prisms and then loaded in direct tension. Strand rupture failures indicated that that lap length for 0.5-inch (12.7-mm)-diameter strands is approximately 18 inches (457 mm), and the lap length for 0.6-inch (15.2-mm)-diameter strands is approximately 26 inches (660 mm).

Steinberg and Lubbers reported the results of pullout tests of 0.5-inch (12.7-mm)-diameter standard and oversize prestressing strands embedded 12, 18, and 24 inches (305, 457, and 610 mm) in UHPC. [204,205] In comparison with conventional concrete having compressive strengths less than 4.0 ksi (28 MPa), the UHPC had higher bond strengths. The results indicated that the strand strength was developed in less than 12 inches (25.4 mm).

Based on tests with 0.5-inch (12.7-mm)-diameter seven-wire strands, Hegger et al. showed that the minimum cover and minimum clear spacing to prevent splitting in UHPC could be reduced to 1.5d and 2.0d where d is the strand diameter. [166] This is less than required by the German DIN 1045-01 for conventional concrete. [159] In other tests, a concrete cover less than 2.5d led to splitting cracks. [206] The authors recommended a minimum cover of 2.5d and a minimum clear spacing of 2.0d. [202]

Structural Analysis

Chen and Graybeal reported the results of a research program to develop finite element analysis modeling techniques applicable to UHPC structural components. [207] The mechanical properties used in the modeling are given in table 12.

Results of the analysis using the values given in table 12 compared favorably with values measured during tests on an I-girder and a pi-girder. [208,209]

Table 12. UHPC properties used in finite element modeling

	Value	
Property	English Units	Metric Units
Unit Weight	160 lb/ft2	2,565 kg/m2
Compressive Strength	29 ksi	200 MPa
Modulus of Elasticity	7,650 to 8,000 ksi	53 to 55 GPa
Poisson's Ratio	0.18	0.18
Post-Cracking Tensile		
Strength	1.4 to 2.3 ksi	9.7 to 15.9 MPa
Ultimate Tensile		
Strain	0.007 to 0.010	0.007 to 0.010

Design Guidelines

The literature search identified the following national recommendations for UHPC: □ Design Guidelines for Ductal Prestressed Concrete Beams (Australia). [210]

- Recommendations for Design and Construction of Ultra High Strength Fiber Reinforced Concrete Structures by the Japan Society of Civil Engineers. [8]
- Ultra High Performance Fibre-Reinforced Concretes, Interim Recommendations prepared by AFGC (French Association of Civil Engineers) and SETRA (French Road and Traffic Government Agency (SETRA-AFGC 2002). [4]

On a more global scale, the Fédération Internationale du Béton (*fib*) Task Group 8.6 is developing recommendations tailored to the design of UHPC structures. [211] The table of contents of the draft version have been published in Walraven. [211]

Design Guidelines in Australia

The Australian guidelines were developed for the design of prestressed concrete beams manufactured using Ductal®. [210] Where possible, a limit state approach consistent with the design requirements of the Australian Standard for Concrete Structures AS3600-1994 was adoptei [212] The design procedures are based on the principles of structural mechanics and the material properties and behavior reported in the literature. Design guidelines are provided for strength, serviceability, and durability.

The material design properties address behavior in compression and tension, modulus of elasticity, density, Poisson's ratio, creep, and shrinkage. Design guidelines are provided for strength in flexure, strength in shear, strength in torsion, flexural crack control at service loads, deflection at service loads, loss of prestress, and anchorage zones.

Theoretical flexural capacity is based on equilibrium of forces and strain compatibility using idealized stress-strain curves in compression and tension for UHPC. A strength reduction factor of 0.8 is used for sections containing bonded reinforcement and 0.7 for sections containing no bonded reinforcement. Ductility is provided by limiting the ratio of neutral axis depth to effective depth to a maximum value of 0.4.

Shear strength of the UHPC in beams is based on limiting the principal tensile stress at the centroidal axis or at the junction of the web and flange to a maximum value based on a section uncracked in flexure. This maximum value is provided in figure 12. When beams contain stirrups or inclined tendons, their contribution to shear resistance may be included in the same way as conventional reinforced concrete design. An equation is provided for the punching shear strength.

$$5.0 + 0.13\sqrt{f_c'} \text{ in SI units}$$

Figure 12. Equation. Shear strength of UHPC beams. [210].

The torsional strength, for a member not containing torsional reinforcement, is taken as the pure torsion required to cause first cracking.

Flexural cracking is controlled by limiting the maximum tensile stress to 870 psi (6.0 MPa) in nonprestressed elements and 1,160 psi (8.0 MPa) in prestressed elements.

Short-term deflections are calculated using conventional procedures for uncracked sections and integration of curvatures for cracked sections. Long-term deflection calculations are based on an age-adjusted effective modulus.

The guidelines suggest that a reliable estimate of prestress losses can be obtained using the age-adjusted effective modulus.

The transfer length of prestressing strands is to be taken between $20d_b$ and $40d_b$ depending on the stress condition being analyzed, where d_b is the strand diameter.

Appendices to the guidelines provide design examples.

This document could provide a template for a similar set of guidelines based on the AASHTO LRFD Bridge Design Specifications. [76]

Design Document from Japan Society of Civil Engineers

The draft recommendations in Recommendations for Design and Construction of Ultra High Strength Fiber Reinforced Concrete Structures (Draft), published by the Japan Society of Civil Engineers in 2006, provide basic principles for design and construction using UHPC. [8] The design values for materials include compressive strength, first cracking strength, tensile strength, stress-strain relationships, modulus of elasticity, Poisson's ratio, thermal characteristics, shrinkage, creep, and fatigue. Other sections address structural safety (strength design), serviceability, fatigue resistance, structural details, prestressed concrete, durability, construction (constituent materials, mix proportions, production, transportation, and inspection), cold-weather concreting, and hot weather concreting. The recommendations build on the Standard Specifications for Concrete Structures prepared by the Japanese Society of Civil Engineers. [213] Both recommendations and extensive commentary are provided.

For flexural design, the use of stress-strain curves rather than an equivalent stress block is recommended. No minimum amount of steel reinforcement is required because the bridging action of the steel fibers provides the strength after cracking.

Shear capacity is calculated as the summation of the shear resistance provided by the matrix, fiber reinforcement, and vertical component of the prestressing force or the shear resistance to diagonal compression failure. The use of shear reinforcement is not recommended. Torsional design is based on the Society's Standard Specifications. An equation is provided for the calculation of punching shear strength.

Serviceability is addressed by checks on stresses, displacements, deformations, vibrations, and other parameters as needed. Verification of fatigue resistance relies on the provisions of the Standard Specifications.

In pretensioned concrete, the clear vertical or horizontal distance between strands may be equal to the strand diameter. A minimum clear cover of 0.8 inches (20 mm) is permitted.

Overall, the document is comprehensive, although it defaults to the Standard Specifications where information is not available to develop different recommendations for UHPC.

AFGC-SETRA Recommendations

The French recommendations are composed of three parts. [4] The first part provides specifications regarding the mechanical properties to be obtained, procedures to be used for placement, and checks and inspection during construction and of the finished product. The second part deals with the design and analysis of UHPC structures and accounts for the participation of fibers, nonprestressed reinforcement, and non-reinforced elements. The third part deals with durability of UHPC.

The first part provides design information for compressive strength, tensile strength, modulus of elasticity, Poisson's ratio, coefficient of thermal expansion, shrinkage, creep, and impact behavior. Mix design, mixing procedures, placement practices, and tests are addressed.

The design methods in the second part are based on the French codes for prestressed and reinforced concrete but take into account the strength provided by the fibers. The recommendations include an orientation coefficient that accounts for the alignment of fibers that may occur during placement. A minimum fiber content and non-brittleness check is also required. The stresses at the serviceability limit state are addressed in the same way as conventional reinforced or prestressed structures. When no prestressing steel or nonprestressed reinforcement is provided, a crack width criterion is used.

For the ultimate flexural strength limit state, the recommendations propose a stress-strain relationship that is linear for the compressive stress range but multilinear in the tensile stress range to account for the effect of the fibers.

At the serviceability limit state for shear, the recommendations use the shear stress limits of the French Code for prestressed concrete. Shear strength is calculated as the summation of the shear resistances provided by the concrete, reinforcement, and fibers.

The components of the third part address water porosity, oxygen permeability, chloride ion diffusion, portlandite content, stability of admixtures, delayed hydration, corrosion of steel fibers, and durability of polymer fibers.

More details on specific topics are provided in nine appendices. Feedback and research resulting from the use of the French recommendations have been summarized by Resplendino. [214]

Summary of Structural Design

Limited testing under flexural or axial loads indicates that the flexural and axial strengths of UHPC members can be calculated with reasonable accuracy if the stress-strain relationships of UHPC are included in the analyses. However, the calculations are more complex than using the simplified approach of a rectangular compressive stress block and zero tensile strength.

The shear strength of UHPC beams containing conventional shear reinforcement and no steel fibers can be predicted using the sectional design method of the AASHTO LRFD Bridge Design Specifications. [76] For UHPC beams with steel fibers and without conventional shear reinforcement, a strength calculation based on the maximum principal tensile stress has been used.

Where design for punching shear is required, the equations in ACI 318 may be used. [162] For shear friction, the available test results need to be compared with the existing specifications.

The limited information available on torsion tests indicates that design could be performed using traditional mechanics of materials approach and limiting the maximum principal tensile stress.

For prestressed concrete, no stress limits or prestress loss values have been established for UHPC. The limited information on transfer length and development length of prestressing strand indicates that the lengths are much shorter in UHPC than in conventional concrete. Similarly, development lengths for deformed bars in tension and lap splices in tension are shorter than for conventional concrete.

For prestress losses, approximate estimates can be made using the modulus of elasticity, creep, and shrinkage data summarized in section 3.

Information on reinforcement details, standard hooks in tension, and development of welded wire reinforcement and shear reinforcement in UHPC members was not identified.

Three countries have developed design guidelines for use with UHPC. Although these documents are not as complete as the AASHTO LRFD Bridge Design Specifications, they do address the major design requirements. [76]

SECTION 5. DURABILITY AND DURABILITY TESTING

Introduction

The use of U H PC i n any infrastructure application requires the U H PC to have adequate resistance to deterioration caused by the environment to which it is exposed. This section reports on the durability of U H PC based on the parameters and tests generally used to determi ne the durability of conventional concrete.

Permeability

In the United States, the permeability of concrete is generally assessed using AASHTO T 277 (ASTM C1202)—Electrical Indication of Concrete's Ability to Resist Chloride Ion Penetration. [215,216] Tests conducted by Graybeal in accordance with ASTM C1202 resulted in values of less than 40 coulombs at 28 days for steam-cured specimens and values of 360 and 76 coulombs at 28 and 56 days, respectively, for untreated specimens. [22] Materials with values less than 100 coulombs are considered to have negligible chloride ion penetrability.

Ahlborn et al. reported rapid chloride permeability values of less than 100 coulombs for both air-cured and heat-treated concretes. [108] Bonneau et al. reported values of 6 to 9 coulombs for two different mixes. [105] Thomas et al. reported values of zero to 19 coulombs at an age of 28 days. [217] Ozyildirim reported values of 19 and 35 coulombs. [45]

Chloride penetration tests in accordance with AASHTO T 259 were also reported by Graybeal. [22,218] This test involves ponding a 3-percent sodium chloride solution on the surface of the concrete for 90 days and then determining the penetration of the chlorides into the concrete. Although there tended to be higher levels of chloride ions near the surface, the amount of chloride that penetrated into the concrete was extremely small.

Different tests for permeability are used in other countries. One measure of chloride penetration is the value of its chloride diffusion coefficient. Reported values are as follows:

- 2×10^{-15} m^2/second in steady-state conditions and 3×10^{-11} m^2/second in non-steady-state conditions. [219]
- 1.3×10^{-13} m^2/second at 28 days. [217]
- 2.3×10^{-13} m^2/second in a non-steady-state condition. [220]

Gao et al. tested the permeability of UHPC using pressure testing. No water leakage occurred when the hydraulic pressure was increased from 14.5 to 232 psi (0.1 to 1.6 MPa) at a rate of 14.5 psi (0.1 MPa) per 8 hours. [91] After removing the pressure, moisture had penetrated 0.11 inches (2.7 mm) into the specimens.

The effects of microcracks induced by loading on chloride penetration have also been investigated. Graybeal examined the penetration of a 15-percent sodium chloride solution into the tension face of a beam. [221] The beam was subjected to 500,000 cycles of repetitive loading over 154 days to a maximum tensile stress 14 percent above the first cracking load. The solution penetrated to a depth of 0.12 inches (3 mm) on the side faces and 0.2 inches (5 mm) on the tensile face of the beam. The steel fibers crossing crack planes did not show any visible signs of section loss or tensile failure.

Aarup loaded small reinforced beams with a cover to the reinforcement of 0.4 inches (10 mm) to produce various levels of bending stresses. [23] Over a period of 4 years, during which the beams were repeatedly exposed to a salt solution for 2 days and dried for 5 days, no correlation between loading of the beams and chloride diffusion was observed and no corrosion occurred. Measured diffusion coefficients for unloaded and loaded beams ranged from 2×10^{-14} to 1×10^{-15} m^2/second.

Charron et al. reported the results of permeability tests on UHPC specimens previously subjected to various levels of tensile deformation. [222] Based on the test results, the maximum residual tensile strain whereby the water permeability remained low was determined to be 0.13 percent.

Freeze-Thaw Resistance

The standard test for freeze-thaw resistance in the United States is AASHTO T 161 (ASTM C666)—Resistance of Concrete to Rapid Freezing and Thawing. [223, 124] AASHTO T 161 has two procedures. Procedure A involves rapid freezing and thawing in water while Procedure B involves rapid freezing in air and thawing in water. Tests of UHPC beginning 5 to 6 weeks after casting and using Procedure A were reported by Graybeal. [22] Specimens subjected to steam curing prior to testing and untreated specimens showed very little deterioration throughout 690 cycles of freezing and thawing. The specimens that were untreated continued to hydrate and gain strength during the testing sequence.

The ability of conventional concrete to resist freeze-thaw damage can also be assessed by measuring certain parameters of its air-void system. Air-void analyses of UHPC reported by Graybeal are shown in table 13. [22]

Table 13. Air-void system parameters

	Value	
Parameter	inches	mm
Voids	2.0 to 7.6/inches	0.08 to 0.30/mm
Specific surface	250 to 405 inches2/inches3	9.8 to 15.9 mm2/mm3
Spacing factor	0.009 to 0.027 inches	0.23 to 0.69 mm

Despite having an air-void system that might not be suitable with conventional concrete, the U H PC performed adequately in freeze-thaw testing.

Bonneau et al. reported that the durability factor of three different mixes was equal to or greater than 100 when tested using A STM C666 Procedure A. [105]

Acker and Behloul reported tests with 400 cycles of freezing and thawing that showed no degradation. [130] Similar results were obtained by A hl born et al. and Pi érard et al. [108,220]

M agureanu et al. reported that U HPC samples displayed higher values for compressive strength, static modulus of elasticity, and dynamic modulus of elasticity after 1,098 freeze-thaw cycles compared with control specimens. [69]

Based on their research, MCiller et al. concluded that UHPC mixes show an extremely high freeze-thaw resistance to water with or without deicing salts. [224] They attributed this to the very low moisture uptake by the UHPC.

Scaling Resistance

The standard test for evaluation of scaling resistance in the United States is ASTM C672— Scaling Resistance of Concrete Surfaces Exposed to Deicing Chemicals. [225] In this test, the surface is exposed to a salt solution and subjected to daily freeze-thaw cycles. Generally, 50 cycles are sufficient to evaluate a surface. Graybeal reported that after 215 cycles, no surface scaling of UHPC specimens had occurred. [22] Bonneau et al. reported very low amounts of scaling for three mixes after 50 cycles. [105]

Schmidt et al. reported a scaling rate of 100 g/m^2 (3 oz/yd^2) after 56 cycles of freezing and thawing compared with the normal acceptance limit for their test of 1,500 g/m^2 (44 oz/yd^2) after 28 cycles. [115] Measurements of sound velocities showed no internal damage from freeze-thaw testing. Specimens that received no heat treatment showed a higher freeze-thaw resistance compared with heat-treated specimens.

Cwirzen et al. examined the effect of heat treatment on the durability of UHPC. [226] The test results for specimens without steel fibers showed low surface-scaling values after 56 freeze-thaw cycles in all specimens. After 150 freeze-thaw cycles, the heat-treated specimens showed an increase in surface scaling. The relative dynamic modulus of the heat-treated specimens dropped below 50 percent after 200 cycles, whereas the non-heat-treated specimens showed a very small change. The presence of steel fibers restrained the internal damage but caused higher surface scaling.

Carbonation

Carbonation of concrete is a process by which carbon dioxide from the atmosphere penetrates the concrete and reacts with various hydration products. Depth of carbonation is typically measured by applying phenolphthalein solution to the surface of the concrete and measuring the depth of the color change. [227]

Small-scale beams of UHPC placed in a carbonation chamber and subjected to flows of 5- or 100-percent carbon dioxide showed no signs of carbonation after 2 years. [219] On the other hand, MCiller et al. reported that mechanically induced microcracks were observed to be partly or completely filled by carbonation. [224]

Piérard et al. reported a carbonation depth of 0.006 to 0.008 inches (1.5 to 2.0 mm) after a 1-year exposure to a 1-percent CO_2 atmosphere. [220] The duration of the test is generally limited to 56 days.

Abrasion Resistance

Graybeal reported tests for the abrasion resistance of UHPC. [22] The tests were conducted in accordance with ASTM C944—Abrasion Resistance of Concrete or Mortar Surfaces by the Rotating-Cutter Method on 6-inch (152-mm)-diameter cylinders that were cured using one of four curing methods. [228] Three concrete surfaces were used—cast against a steel mold, sand blasted, and ground. The double test load was used. The results clearly indicated much higher abrasion resistance of steam-cured specimens compared with untreated specimens. For the steam-cured specimens, the surfaces cast against the steel mold had higher abrasion resistance than the sand-blasted or ground surfaces.

Sulfate Resistance

Piérard reported no deterioration of UHPC when immersed in sodium sulfate solution for 500 days. [220]

Resistance to Alkali-Silica Reactivity

Various tests to determine the resistance of concrete to alkali-silica reactivity (ASR) are available. [229] ASTM C1260 contains a test procedure that accelerates any ASR reaction and can be accomplished in 16 days. [230] Using a version of this test modified to allow for steam curing, Graybeal reported levels of expansion that were an order of magnitude below the threshold value for innocuous behavior. [22] He concluded that there should be no concern about ASR with the UHPC that was tested. He noted that free water must be present for ASR to occur. With the low permeability of UHPC, it is unlikely that free water would be present.

Marine Exposure

Three series of UHPC mixtures were placed in a marine exposure site at Treat Island, ME. [217] The exposure conditions included 20-ft (6-m) tides and more than 100 freeze-thaw cycles per year. After 5 to 15 years of exposure and more than 1,500 cycles of freezing and thawing in some cases, there was no evidence of deterioration or degradation of mechanical properties. The depth of chloride penetration was much lower than observed for typical HPC in the same environment.

Fire Resistance

Behloul et al. have reported information related to the fire resistance of UHPC made with Ductal-AF®. [122] Ductal-AF® is specially formulated to have fire resistance. Published information includes the change in compressive strength, tensile strength, modulus of elasticity, thermal conductivity, specific heat, and coefficient of thermal expansion for specimens subjected to temperatures between 68 and 1,112 °F (20 and 600 °C).

Fire tests according to ISO 834 were also conducted on columns and beams using both loaded and unloaded specimens. [231,122] Some specimens were steam cured while others were not. The authors reported that the results were very positive compared with conventional concrete when using the French rules for fire safety. One feature was the lack of spalling that occurs with conventional concrete. This facilitated the use of thermal modeling to predict the behavior.

Heinz et al. reported the fire resistance of UHPC 3.9-inch (100-mm)-diameter cylinders and 4.7- by 9.4-inch (120- by 240-mm) columns under load. [232] The concretes included either steel fibers or a combination of steel and polypropylene fibers. At an age of 24 hours, the specimens were heat treated in water at 194 °F (90 °C) for 24 hours. Testing followed the time-temperature curve of German standard DIN 4102-2. [233] The cylinders without polypropylene fibers exhibited spalling after a few minutes. After 90 minutes, the sample was destroyed beyond recognition. In contrast, cylinders containing 0.66 percent by volume of polypropylene fibers showed no signs of spalling. However, cracks with widths of 0.012 to 0.02 inches (0.3 to 0.5 mm) were present over the whole surface of the cylinders. In testing the columns, spalling occurred after about 11 minutes. The initial period of spalling was followed by a dormant period with no further destruction until fracture of the specimens. The authors concluded that a UHPC with 3.05 percent by volume of steel fibers and 0.60 percent by volume of polypropylene provided the best results. The effects of elevated temperatures on the residual compressive strength and modulus of elasticity were also reported by Way and Wille. [234]

Hosser et al. also conducted tests to evaluate which combinations of protective lining and polypropylene fiber content were able to minimize spalling under fire exposure. [235] They also measured thermal conductivity and specific heat.

Aarup reported that the behavior of UHPC 1 week after fire tests was better than for conventional concrete. [23] One reason stated for the improved performance was that the UHPC had a very high silica fume content and negligible calcium hydroxide content. A literature review of the behavior of UHPC at elevated temperatures has been prepared by Pimienta et al. [236]

Summary of Durability Properties

The dense matrix of UHPC prevents deleterious solutions from penetrating into the matrix, and so the mechanisms that can cause conventional concrete to deteriorate are not present. Consequently, durability properties, as measured by permeability tests, freeze-thaw tests, scaling tests, abrasion tests, resistance to ASR, and carbonation, are significantly better than those of conventional concrete. For fire resistance, it appears that a special formulation may be necessary.

Section 6. Actual and Potential Applications

This section describes specific applications of UHPC in infrastructure projects. Separate sections contain descriptions of the applications in North America (United States and Canada), Europe, and Asia/Australasia. Potential applications described in the literature are also presented.

North America

Table 14 provides a list of the applications in the United States and Canada.

Table 14. UHPC applications in North America

Name	Country	Year	Application	Reference (First Author)
Mars Hill Bridge, Wapello County, IA	United States	2006	Three 45-in.-deep bulb-tee beams	Bierwagon [237] Endicott [238]
Route 624 over Cat Point Creek, Richmond County, VA	United States	2008	Five 45-inch-deep bulb-tee girders	Ozyildirim [45]
Jakway Park Bridge,	United	2008	Three 33-inch-deep pi-	Keierleber [239]

Name	Country	Year	Application	Reference (First Author)
Buchanan County, IA	States		shaped girders	
State Route 31 over Canandaigua Outlet, Lyons, NY	United States	2009	Joints between deck bulb tees	Shutt [240]
State Route 23 over Otego Creek, Oneonta, NY	United States	2009	Joints between full-depth deck panels	Royce [241]
Little Cedar Creek, Wapello County, IA	United States	2011	Fourteen 8-inch-deep waffle deck panels	Moore [242]
Fingerboard Road Bridge over Staten Island Expressway, NY	United States	2011 to 2012	Joints between deck bulb tees	Royce [241]
State Route 248 over Bennett Creek, NY	United States	2011	Joints between deck bulb tees	Royce [237]
U.S. Route 30 over Burnt River and UPRR bridge, Oregon	United States	2011	Haunch and shear connectors and transverse joints	Bornstedt [243]
U.S. Route 6 over Keg Creek, Pottawatomie County, IA	United States	2011	Longitudinal and transverse joints between beams	Graybeal [63]
Ramapo River Bridge, Sloatsburg, NY	United States	2011	Joints between full-depth deck panels	Anon [244]
State Route 42 Bridges (2) near Lexington, NY	United States	2012	Joints between full-depth deck panels and shear pockets	Anon [244]
State Route 31 over Putnam Brook near Weedsport, NY	United States	2012	Joints between full-depth deck panels	Anon [244]
I-690 Bridges (2) over Peat Street near Syracuse, NY	United States	2012	Joints between full-depth deck panels	Anon [244]
I-690 Bridges (2) over Crouse Avenue near Syracuse, NY	United States	2012	Joints between full-depth deck panels	Anon [244]
I-481 Bridge over Kirkville Road near Syracuse, NY	United States	2012	Joints between full-depth deck panels	Anon [244]
Windham Bridge over BNSF Railroad on U.S. Route 87 near Moccasin, Montana	United States	2012	Joints between full-depth deck panels and shear connections to beams	Anon [244]
Sherbrooke Pedestrian Overpass, Quebec	Canada	1997	Precast, post-tensioned space truss	Blaise [2]
Highway 11 over CN Railway at Rainy Lake, Ontario	Canada	2006	Joints between precast panels and shear connector panels	Perry [245]
Glenmore/Legsby Pedestrian Bridge, Calgary	Canada	2007	Precast, post-tensioned tee-section	Perry [246]
Highway 11/17, Sunshine Creek, Ontario	Canada	2007	Joint fill between adjacent box beams and between precast curbs	Graybeal [139]

Table 14. (Continued)

Name	Country	Year	Application	Reference (First Author)
Highway 17, Hawk Lake, Ontario	Canada	2007 to 2008	Joint fill between adjacent box beams and between precast curbs	Graybeal [139]
Sanderling Drive Pedestrian Overpass, Calgary	Canada	2008	Tee section drop-in girder	Anon [244]
Highway 105 over Buller Creek, Ontario	Canada	2009	Joint fill between adjacent box beams and between precast curbs	Graybeal [139]
Highway 71 over Log River, Ontario	Canada	2009	Joint fill between adjacent box beams and between precast curbs	Graybeal [139]
Route 17 over Eagle River, Ontario	Canada	2010	Joint fill between adjacent box beams and between precast curbs and to establish live load continuity	Graybeal [63,139]
La Vallee River Bridge, Ontario	Canada	2010	Joint fill between adjacent box beams and between precast curbs	Graybeal [139]
Highway 105 over Wabigoon River, Ontario	Canada	2010	Joint fill between adjacent box beams and between precast curbs	Graybeal [139]
Highway 105 over the Chukuni River, Ontario	Canada	2010	Shear connector pockets and panel joints	Graybeal [139]
Steel River Bridge on Highway 17, Ontario	Canada	2010	Shear connector pockets and panel joints	Anon [244]
Mathers Creek Bridge on Highway 71, Ontario	Canada	2010	Joint fill between adjacent box beams and between precast curbs	Anon [244]
Noden Causeway on Highway 11, Ontario	Canada	2010 to 2013	Joint fill between adjacent precast panels	Anon [244]
Highway 17 over Current River, Ontario	Canada	2011	Joints between precast curbs	Perry [247]
Mackenzie River Bridges (2) on Highway 11/17, Ontario	Canada	2011	Shear connector pockets and panel joints	Anon [244]
Wabigoon River Bridge on Highway 605, Ontario	Canada	2011	Shear connector pockets and panel joints	Anon [244]

Name	Country	Year	Application	Reference (First Author)
Whiteman Creek Bridge on Highway 24, Ontario	Canada	2011	Shear pockets and longitudinal and transverse joints between precast panels. Connections between H-piles and precast abutments	Young [248,249]
Shashawanda Creek Bridge, Ontario	Canada	2011	Shear connector pockets and longitudinal and transverse joints between precast panels	Anon [244]
Hodder Ave Overpass over Highway 11/17, Ontario	Canada	2012	Joint fill between adjacent box beams and between precast curbs	Anon [244]
Hawkeye Creek Bridge on Highway 589, Ontario	Canada	2012	Joint fill between adjacent box beams and between precast curbs	Anon [244]
Hawkeye Creek Tributary Bridge on Highway 589, Ontario	Canada	2012	Joint fill between adjacent box beams and between precast curbs	Anon [244]
Black River Bridge on Highway 17, Ontario	Canada	2012	Joint fill between adjacent box beams and between precast curbs	Anon [244]
Beaver Creek Bridge on Highway 594, Ontario	Canada	2012	Joint fill between adjacent box beams and between precast curbs	Anon [244]
Middle Lake Bridge on Highway 17A, Ontario	Canada	2012	Joint fill between precast curbs and precast approach slabs	Anon [244]
Jackpine River Bridge on Highway 17, Ontario	Canada	20131	Joint fill between adjacent box beams and between precast curbs	Young2
Bug River Bridge on Highway 105, Ontario	Canada	20131	Joint fill between adjacent box beams and between precast curbs	Young2
Beaver Creek Bridge on Highway 17, Ontario	Canada	20131	Joint fill between adjacent box beams and between precast curbs	Young2

Table 14. (Continued)

Name	Country	Year	Application	Reference (First Author)
Sturgeon River Bridge on Highway 11, Ontario	Canada	2013[1]	Joint fill between adjacent box beams and between precast curbs	Young[2]
Blackwater River Bridge on Highway 11, Ontario	Canada	2013[1]	Joint fill between adjacent box beams and between precast curbs	Young[2]
Nugget Creek Bridge on Highway 17, Ontario	Canada	2013[1]	Joint fill between adjacent box beams and between precast curbs	Young[2]
Little Wabigoon Bridge on Highway 17, Ontario	Canada	2013[1]	Joint fill between adjacent box beams and between precast curbs	Young[2]
Melgund Creek Bridge on Highway 17, Ontario	Canada	2013[1]	Joint fill between adjacent box beams and between precast curbs	Young[2]
McCauley Creek Bridge on Highway 11, Ontario	Canada	2013[1]	Joint fill between adjacent box beams and between precast curbs	Young[2]
Little Pic River Bridge on Highway 17, Ontario	Canada	2013[1]	Shear connector pockets and panel joints	Young[2]
Jackfish River Bridge on Highway 17, Ontario	Canada	2013[1]	Shear connector pockets and panel joints	Young[2]
Westminster Drive, Ontario	Canada	2014[1]	Longitudinal joints to connect superstructure modules.	Young[2]

[1] Projected construction date.

[2.] W. Young to B. Graybeal, personal email communication, December 21, 2012.

The first highway bridge constructed in North America was the Mars Hill bridge in Wapello County, IA. [238] The simple single-span bridge, as shown in figure 13, comprises three 110-ft (33.5-m)-long precast, prestressed concrete modified 45-inch (1.14-m)-deep Iowa bulb-tee beams topped with a cast-in-place concrete bridge deck. Each beam contained forty-seven 0.6-inch (15.2-mm)-diameter, low-relaxation prestressing strands and no shear reinforcement.

Figure 13. Photo. Mars Hill Bridge, Wapello County, IA.

One span of the 10 spans of the Route 624 bridge over Cat Point Creek in Richmond County, VA, was built using UHPC. [45] (See figure 14.) Bulb-tees with a depth of 45 inches (1.14 m) and a length of 81 ft 6 inches (24.8 m) were used. The specified compressive strengths were 12.0 ksi (83 MPa) at release of the strands and 23.0 ksi (159 MPa) for design. The beams did not contain any nonprestressed shear reinforcement.

Figure 14. Photo. Route 64 over Cat Point Creek, Richmond County, VA.

Following extensive research and testing by FHWA, a UHPC bridge using pi-shaped girders was constructed in Buchanan County, IA, in 2008. [239,250] (See figure 15.) The shape is named after the Greek letter . The cross section, shown in figure 16, is similar to a double-tee section but with bottom flanges on the outside of each web. Three pi-girders were used in the central 51-ft 4-inch (15.6-m)-long center span of the three-span bridge.

Figure 15. Photo. Jakway Park Bridge, Buchanan County, IA.

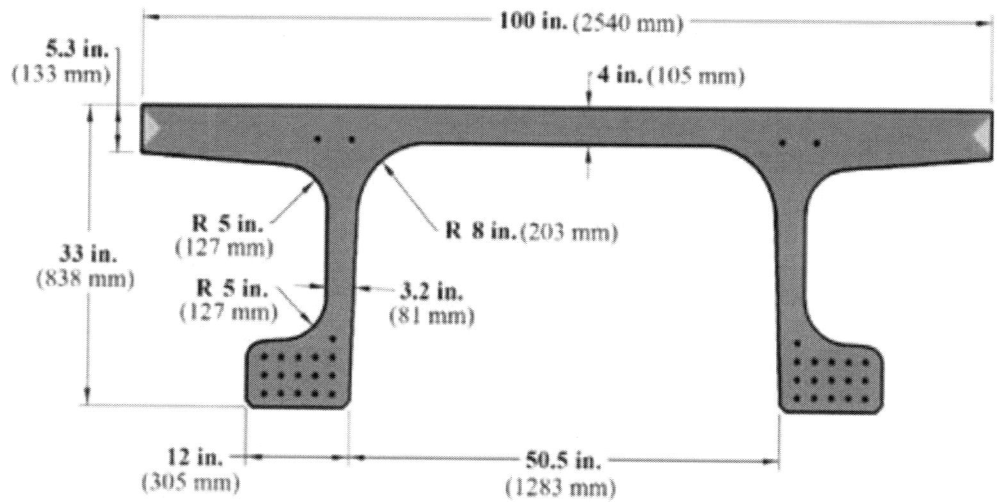

Figure 16. Illustration. Cross section of pi-shaped girder.

In New York State, several bridges have been built using field-cast UHPC to create connections between adjacent precast concrete elements. [241] (See figure 17.) These applications take advantage of the short development lengths that can be used for splice lengths of nonprestressed reinforcement in UHPC. The same technique was used on the transverse joints over the piers of the Keg Creek Bridge, IA, to establish continuity for live load and in the longitudinal joints between deck panels. The use of UHPC in the construction of connections is described by Graybeal. [251]

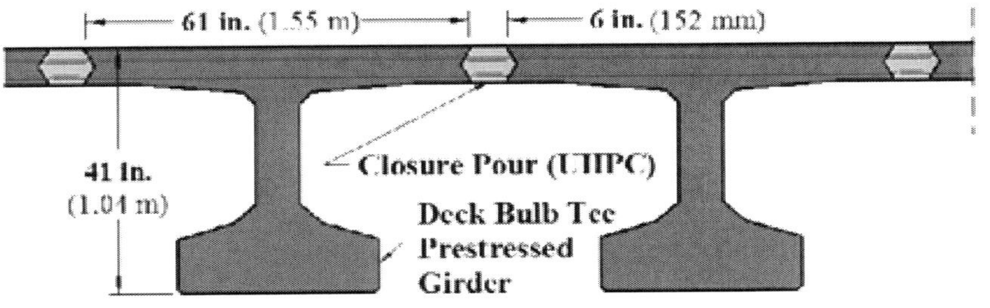

Figure 17. Illustration. Cross section showing CIP UHPC connection between precast beams.

Little Cedar Creek in Wapello County, IA, used 14 UHPC waffle panels for the deck on a 60-ft (18.3-m)-long 33-ft (10.0-m)-wide concrete bridge. [242] The panels were 15 ft by 8 ft by 8 inches deep (4.6 m by 2.4 m by 203 mm deep) at the deepest point, with the waffle squares having a thickness of only 2.5 inches (64 mm). All connections between adjacent panels and from panels to the precast, prestressed concrete beams used UHPC.

The first bridge to use UHPC in Canada was the pedestrian/bikeway bridge in Sherbrooke, Quebec, as shown in figure 18. [2] The structural concept consists of a space truss with a top UHPC chord that serves as the riding surface, two UHPC bottom chords, and truss diagonals that slope in two directions. Each diagonal consists of UHPC confined in 6-inch (152-mm)-diameter stainless steel tubes. The bridge was constructed from six prefabricated match-cast segments with two half-spans assembled prior to erection across the river to create a 197-ft (60-m)-long span.

Source: Lafarge.

Figure 18. Photo. Pedestrian bridge, Sherbrooke, Quebec, Canada.

Other bridges in Canada that have used UHPC are listed in table 14. The applications include longitudinal and transverse joints between precast components, shear connector pockets between beams and slabs, and a precast post-tensioned tee section for a pedestrian bridge. See Figure 19. Most of the applications have been in Ontario with leadership by the Ministry of Transportation.

Source: Lafarge.

Figure 19. Photo. Glenmore/Legsby pedestrian bridge, Calgary, Alberta, Canada.

Europe

UHPC has been used in bridges in Austria, Croatia, France, Germany, Italy, the Netherlands, Slovenia, and Switzerland as listed in table 15.

The Bourg-Les-Valence bridges in France are claimed to be the first UHPC road bridges. [256] Each bridge consists of two spans made continuous with a CIP UHPC connection between spans. The cross section consists of five spliced pretensioned beams that resemble a double-tee with the addition of bottom flanges similar to a pi-shaped section. Beam lengths are 67.3 and 73.8 ft (20.5 and 22.5 m). The only nonprestressed reinforcement is provided where the components are joined together longitudinally or transversely and at locations of attachments. UHPC was used in the longitudinal joints between beams.

The PS34 Overpass on the A51 motorway in France is a precast, post-tensioned, single-cell box girder bridge with a length of 155.5 ft (47.4 m). [257] The cross section has a constant depth of 63 inches (1.60 m), a top slab thickness of 5.5 inches (140 mm), and web and bottom slab thickness of 4.7 inches (120 mm). The bridge is post tensioned longitudinally with six external tendons.

The St. Pierre La Cour bridge in France consists of 10 UHPC precast, prestressed concrete I-beams spaced at 55-inch (1.395-m) centers with a simple span length of 62.3 ft (19 m). [257] The deck consists of 1-inch (25-mm)-thick UHPC precast panels and an 8-inch (200-mm)-thick CIP deck.

According to Fehling, the first UHPC bridges in Germany were built in Niestetal near Kassel with span lengths of 23.0, 29.5, and 39.4 ft (7, 9, and 12 m). [260] The longest span used a shallow trough section and was post tensioned. The other two spans used a pi-shaped section and were pretensioned. Two other bridges using the pi-shaped cross-section were built near Friedberg and Weinheim with span lengths of 39.4 and 59.0 ft (12 and 18 m), respectively.

Table 15. UHPC applications in Europe

Name	Country	Year	Application	Reference (First Author)
WILD bridge,Völkermarkt	Austria	2010	Arch bridge with five straight chords	Freytag [252] Hecht [253]
Bakar bridge	Croatia	—	Arch bridge	Čandrlić [254]
Sermaises footbridge	France	—	U-shaped footbridge with a 30-min fire rating	Behloul [255]
Bourg-Les-Valence overpass bridges [2]	France	2001	Pi-shaped beams (double tee)	Hajar [256]
PS 34 overpass on the A51 Campenon Bernard	France	2005	Precast, posttensioned segmental single cell box girder	Resplendino [257]
Sainte Pierre La Cour bridge, Mayenne	France	2005	Precast, prestressed Ibeams and deck panels	Resplendino [257]
Pinel bridge, Rouen	France	2007	Prestressed beams	de Matteis [258]
Pont du Diable footbridge	France	2008	Prestressed beams and deck to form a U-shape	Behloul [259]
TGV East High Speed Line, aqueduct	France	—	Post-tensioned U-shape	Resplendino [214]
Angels footbridge, Herault	France	—	221-ft span, 5.9 ftdeep section	Resplendino [214]
Name	Country	Year	Application	Reference (First Author)
Pedestrian/cycle track Niestetal	Germany	—	Post-tensioned trough section	Fehling [260]
Gaertnerplatz bridge, Kassel	Germany	2007	Variable depth space truss	Fehling [260,261]
Obertiefenbach	Germany	2007	Waterproofing layer and hinge	Kim [43]
Friedberg	Germany	2007	Pi-shaped beam	Fehling [260]
Weinheim	Germany	2007	Pi-shaped beam	Fehling [260]
—	Italy	—	Bridge	Meda [262]
Rehabilitation of orthotropic bridge deck, Caland	Netherlands	—	Toppings and deck panels	Buitelaar [263] Yuguang [264]
Kaag bridges, Sassenheim	Netherlands	2002	Deck panels	Kaptijn [265]
Log Cezsoski bridge	Slovenia	2009	Bridge deck overlay	Sajna [266]
Luaterbrunnen footbridge	Switzerland	—	Flooring	Resplendino [267]
Single span road bridge	Switzerland	2004	Rehabilitation and widening of a bridge deck	Brühwiler [268]
Crash barrier repair	Switzerland	2006	Protective surface layer	
Bridge pier repair	Switzerland	2007	Precast panels for a protective layer	Brühwiler[268]
Various	Various	—	Repair and strengthening -	Resplendino [214]

— Construction date is unknown.

The Gaertnerplatz bridge, a pedestrian/bicycle bridge across the Fulda River in Kassel, Germany, is a six-span structure with a total length of 437 ft (133.2 m) and a main span of 118 ft (36 m). [261] The structural system is a variable-depth space truss consisting of two top UHPC chords and a single bottom tubular steel chord. The diagonal tubular steel chords are inclined both longitudinally and transversely. The deck spans between and cantilevers beyond the two top chords for a total width of 16.4 ft (5 m). Its thickness varies from 3.1 to 3.9 inches (80 to 100 mm). The deck is glued to the top chords.

In Slovenia, a bridge deck was overlaid with 1 to 1.2 inches (25 to 30 mm) of UHPC. [266] An inspection 2 years after installation showed no damage, cracks, or spalling. Applications in Switzerland include rehabilitation and widening of an existing bridge, protection layers to repair a crash barrier and bridge piers, and flooring for a footbridge. [267,268]

Asia and Australasia

UHPC applications for highway infrastructure in Australia, Japan, Malaysia, New Zealand, and South Korea are listed in table 16. Descriptions of some of these bridges are provided below.

Table 16. UHPC applications in Asia and Australia

Name	Country	Year	Application	Reference (First Author)
Shepherds Creek Road bridge, New South Wales	Australia	2005	Precast, pretensioned I-beams	Rebentrost [269] Anon [270] Cavill [271]
Yarra River bridge	Australia	2008 to 2009	Noise barrier protection panels	Anon [272]
Kuyshu Expressway bridge	Japan	—	—	Okuma [273]
Riverside Senshu footbridge, Nagaoka-shi	Japan	—	Three-span continuous structure	Matsubara [274]
Sakata-Mirai footbridge, Sakata	Japan	2002	Post-tensioned box girder	Rebentrost [269] Resplendino [275] Tanaka [276]
Akakura Onsen Yukemuri pedestrian bridge	Japan	2004	Prestressed U-shaped girder	Tanaka [276]
Yamagata	Japan	2004	Box girder	Rebentrost [269]
Tahara bridge Aichi	Japan	2004	Box girder	Rebentrost [269]
Horikoshi Highway C-ramp Fukuoka	Japan	2005	Composite I-girder	Rebentrost [269] Tanaka [276]
Keio University footbridge,Tokyo	Japan	2005	Pretensioned slab	Rebentrost [269] Tanaka [276]
Name	Country	Year	Application	Reference (First Author)
Torisaka, River Highway bridge, Hokkaido	Japan	2006	Launching nose	Rebentrost [269]

Name	Country	Year	Application	Reference (First Author)
Toyota City Gymnasium footbridge, Aichi	Japan	2007	Box girder	Tanaka [276]
Sankin-ike footbridge, Fukuoka	Japan	2007	Box girder	Rebentrost [269]
Hikita pedestrian bridge, Tottori	Japan	2007	U-shaped girder	Rebentrost [269]
Haneda Airport Runway D, Tokyo	Japan	2007	Precast, pretensioned slabs	Rebentrost [269] Tanaka [276]
Mikaneike footbridge. Fukuoka	Japan	2007	U-shaped girder	Musha [277]
Kobe Sanda premium outlet footbridge	Japan	2008	U-shaped girder	Tanaka [276]
Akasaka Yogenzaka footbridge	Japan	2009	U-shaped girder	Tanaka [276]
Torisalogawa bridge	Japan	2006	Box girder	Tanaka [276]
Tokyo Monorail	Japan	2007	U-girder upside down	Tanaka [276]
GSE bridge Tokyo Airport	Japan	2008	U-girders	Tanaka [276]
Kampung Linsum bridge Rantau, Negeri Seremban	Malaysia	—	U-beam	Lei [278] Voo [150]
Sungai Muar bridge	Malaysia	—	Curved saddles for cable stays	Resplendino [275]
Papatoetoe footbridge	New Zealand	2005	Pi-beam	Anon [279]
Five pedestrian bridges, Auckland	New Zealand	2006 to 2007	Precast, post-tensioned Pi-girder	Rebentrost [269] Anon [279]
Seonyu Sunyudo footbridge, Seoul (Peace Bridge)	South Korea	2002	Precast, post-tensioned pi-section	Rebentrost [269] Resplendino [275]
Office pedestrian bridge	South Korea	2009	Cable-stayed bridge	Kim [9]

— Data are unknown.

The Shepherds Creek bridge in Australia is a single 49-ft (15-m)-span bridge with a 16-degree skew. [269,271] The superstructure consists of sixteen 23.6-inch (600-mm)-deep precast, prestressed UHPC beams spaced at 51 inch (1.3 m) centers. These support 1-inch (25-mm)-thick precast UHPC panels and a 6.7-inch (170-mm)-thick CIP reinforced concrete deck.

Numerous bridges, as listed in table 16, have been constructed in Japan beginning with the Sakata-Mirai bridge in 2002. [269,276] (See figure 20.) This footbridge consists of pretensioned box girder segments that were post tensioned together to form a single span of 161 ft (49.2 m).

Most of the UHPC footbridges in Japan consist of precast segmental U-beams with a separate top slab that is integrally connected to the U-beam. The U-beam segments are connected longitudinally with a CIP joint and post tensioning.

The Horikoshi Highway C-Ramp bridge was Japan's first highway bridge using UHPC. [276] The composite girder bridge is composed of four pretensioned UHPC I-shaped girders and a conventional CIP concrete deck. The use of UHPC in the girders allowed reduction of the number of girders from 11 to 4. The weight of each girder was less than it would have been with conventional concrete, allowing the use of a smaller crane. The overall weight of the bridge was reduced by 30 percent.

Source: Lafarge.

Figure 20. Photo. Sakata-Mirai bridge, Sakata, Japan.

Source: Rualt Philippe.

Figure 21. Photo. Footbridge of Peace, Seoul, South Korea.

The Toyota Gymnasium footbridge is a two-cell segmental box girder using match-cast segments and dry joints with epoxy. To overcome the shortening caused by autogenous shrinkage of the lead segment before casting the next segment, a steel plate was used at the end of the lead segment and becomes the end form for the new segment. [276]

The construction of Runway D at Tokyo's Haneda International Airport used 9.8-inch (250-mm)-deep UHPC panels spanning between longitudinal steel girders above the Tamar River. [276] The panels consist of ribs supporting a slab with a minimum thickness of 3 inches (75 mm). This reduced the dead load of the slab by about 56 percent compared with conventional concrete. Approximately 6,900 panels were produced for this application.

The Sunyudo (Peace) footbridge in South Korea is an arch bridge with a main span of 394 ft (120 m). [275] (See figure 21.) It is built from six precast, post-tensioned pi-shaped sections 4.3 ft (1.30 m) deep. The upper flange is a ribbed slab 1.19 inches (30 mm) thick with transverse prestressing. The webs of the pi-shaped section are 6.35 inches (160 mm) thick and inclined outward at the bottom. The six precast sections are post tensioned together by tendons located in the upper and lower haunches of the section. This bridge is the longest span UHPC bridge in the world.

Realized and Potential Security Applications

Significant research and development efforts have also occurred with regard to the potential security applications afforded by UHPC. Infrastructure security can be a critical consideration, thus leading to opportunities to use UHPC components either as barrier protection systems or as inherent portions of the critical infrastructure. A state-of-the-art report on fiber-reinforced UHPC with a focus on security applications was completed in 2010. [280]

Research on the mechanical properties of UHPC when subjected to high strain rate loading has been completed by Parent et al., Ngo et al., Millard et al., Habel and Gauvreau, and Millon et al. (See references 281, 282, 283, 284, and 285.) Blast resistance testing has been reported by Wu et al., Ngo et al., and Rebentrost and Wight. (See references 286, 282, 287, and 288.) Penetration resistance tests have been reported by Rebentrost and Wight [287,288] and by Nöldgen et al. [289]

Other Potential Applications

This section identifies other potential applications found during the literature search.

Almansour and Lounis compared the design of a prestressed concrete girder bridge using either UHPC or HPC in the girders. [290] The design of the UHPC bridge was based on a combination of the Canadian Highway Bridge Design Code (CHBDC) and the AFGC-IR-02. [291,4] The design of the HPC bridge was based only on the CHBDC. Both bridges had a span length of 147.6 ft (45 m). Five girders with a depth of 63 inches (1,600 mm) were required for the HPC bridge, and only four girders with a depth of either 35.4 inches (900 mm) or 47.2 inches (1200 mm) were required for the UHPC bridge. The 47.2-inch (1,200-mm)-deep girders represented a conservative design, whereas the shallower sections required more prestressing strands. An optimum solution would be a girder with a depth between 35 and 47 inches (900 and 1,200 mm).

The design of a pilot project for a 39-ft (12-m) span pedestrian bridge using composite steel-concrete construction was reported by Jungwith et al. [188]

Obata et al. examined the use of prefabricated UHPC panels 1.2 inches (30 mm) thick as an overlay for asphalt pavement. [292] The panels were bonded to the asphalt using a grout. About 517 ft^2 (48 m^2) of test pavement was constructed at a test track in Japan using different construction bonding procedures. No cracks were observed before load testing began. Delaminations occurred and increased with the number of wheel passes in some test sections. The authors concluded that early opening of the pavement to traffic is possible with the use of high-strength fast-curing grout.

Oesterlee et al. performed finite element analyses of a conceptual bridge girder using UHPC as an overlay material in place of a conventional waterproofing membrane. [293] The structural response under combined loading from restrained shrinkage and traffic loads showed stresses close to the elastic tensile strength of the UHPC overlay where there was a high degree of restraint. The risk of transverse cracking in the overlay was deemed unlikely.

Schafers and Seim described theoretical and experimental investigations into the composite behavior of UHPC decks on timber beams. [294] They conducted shear tests of the glued joint between the UHPC and timber to identify the best adhesives and timber surface preparation methods.

Source: Iowa State University.

Figure 22. Photo. Experimental precast pile made of UHPC.

Using finite element modeling and experimental verification, Toutlemende et al. investigated the possible use of UHPC precast ribbed waffle slabs for a bridge deck. [175] The slabs were pretensioned in the transverse direction and then post tensioned longitudinally before being connected to the longitudinal steel girders. The test results were compared with analytical models. [295]

Vande Voort et al. explored the use of UHPC in H-shaped precast, prestressed concrete piles. [296] They used laboratory tests to verify moment-curvature response. Two piles were successfully driven into clay soils and tested under vertical and lateral loads. (See figure 22.) The impact resistance of UHPC for use in piles was investigated by Leonhardt et al. [127]

Other potential applications that have been investigated are listed in table 17.

Table 17. Other potential applications of UHPC

Application	Reference (First Author)
Drill bits for special foundation engineering	Ibuk [297]
Sewer pipes	Schmidt [298]
Precast spun columns and poles	Adam [299], Müller [300]
Barrier walls	Young [249]
Field-cast thin-bonded overlays	Young [249], Sritharan [301], Shann [302], Schmidt [303], Scheffler [55]
Cable-stayed bridge superstructure	Kim [9], Park [304]
Bridge bearings	Hoffmann [305]
Precast tunnel segments	Randl [306]
Seismic retrofit of bridge columns	Massicotte [307]

SECTION 7. FUTURE DIRECTION

This state-of-the-art report identifies the following four primary characteristics of UHPC that distinguish it from conventional concrete:

- Higher compressive strength.
- Higher tensile strength with ductility.
- Increased durability.
- Higher initial unit cost.

The compressive strength of UHPC makes it an ideal material for use in applications in which compressive stress is the predominant design factor. The ductility in tension allows the tensile strength of UHPC to be considered in both service and strength design for flexure, shear, and torsion. The durability of UHPC makes it an ideal material for use in an outdoor or severe exposure environment. The higher initial unit cost means that its use needs to be optimized for the intended application and that greater attention should be given to life-cycle costs. In addition, specifiers should consider all costs associated with the use of UHPC on a project, not just the material unit cost. In many cases, the use of UHPC may allow a redesign of the structure thus affecting many aspects of the total cost of deploying the structure. For example, the ability to omit shear reinforcement in a beam can result in a savings of both materials and labor that must be considered alongside the increased material costs. Nevertheless, a number of challenges must be overcome to achieve wide-scale implementation in the U.S. highway infrastructure. These are outlined in the following sections.

Owner Issues

One of the primary advantages of UHPC to owners is its long-term durability. As discussed in section 5, the measured durability characteristics far exceed those of conventional concrete. These characteristics should result in structures with a longer service life compared with structures built with conventional concrete, and thus could potentially decreased life-cycle costs. No studies were identified for this report to show that this is the case. When owners began to consider the use of high-strength concrete in bridge beams, a clear case could be made that the initial cost would be less because the number of beams for a given bridge would be reduced. This may not be true with UHPC because the cost differential between conventional concrete and UHPC is much greater than it was between conventional concrete and high-strength concrete. Studies are needed to illustrate the cost benefits of using UHPC for bridges in the United States.

The number of demonstration projects in the United States is limited, with most occurring in only two States. For owners to obtain a reasonable level of comfort in using UHPC, more demonstration projects are needed, and the results need to be disseminated through a variety of channels. These include webinars, in-house seminars, technical symposia, and technical publications. Some of this activity has been ongoing for the past 10 years but more is needed. This is not just for owners but also for bridge designers, contractors, and producers.

There are, however, situations where UHPC can be used to address certain performance issues without a major cost impact. One example is the use of UHPC to fill the connection regions between adjacent prefabricated elements. In this application, the overall cost increment in using UHPC is small because the quantity of material is small. The use of UHPC is reported to eliminate the cracking and leakage that occurs when conventional concretes or grouts are used. At the same time, the use of UHPC can enable the deployment of simplified connection details with shorter discrete reinforcement splice lengths and a reduced number of conflict points.

Design Issues

The literature search identified the following national design and construction recommendations for UHPC:

- Design Guidelines for Ductal Prestressed Concrete Beams (Australia).[210]
- Recommendations for Design and Construction of Ultra High Strength Fiber Reinforced Concrete Structures by the Japan Society of Civil Engineers. [8]
- Ultra High Performance Fibre-Reinforced Concretes, Interim Recommendations prepared by AFGC (French Association of Civil Engineers) and SETRA (French Road and Traffic Government Agency (SETRA-AFGC 2002). [4]

These documents are generally based on the primary document used for bridge design in the individual country. Where sufficient information is not available to support a change or a change is not necessary for UHPC, the documents resort to the provisions of the primary document.

For UHPC to gain greater use in the U.S. highway infrastructure, a design and construction document based on the AASHTO LRFD Bridge Design Specifications and the AASHTO LRFD Bridge Construction Specifications is needed. [76,308] The lack of this document has led to the need to consider each project individually. In most cases, the design has been accepted based on structural tests rather than a rational design basis. A guide specification for construction with UHPC will help owners implement the technology.

Although more research is desirable, it is likely that sufficient information exists today to develop a document addressing the major aspects of structural design according to U.S. practices. These design aspects include material properties, flexural and axial load, tensile load, shear, transfer and development length of prestressing strand, approximate estimates of time-dependent losses based on creep and shrinkage data, some aspects of reinforcement details, and durability. Where information is lacking, the document could use the provisions of the existing bridge specifications. This concept may not immediately result in the most economical design but will generally be conservative. Because several demonstration projects have been completed in the United States, there should be sufficient experience available to identify the necessary provisions in a construction guide specification.

For proper implementation of UHPC, new test procedures that address UHPC are needed for both development of mixes and quality control of the fresh and hardened UHPC. In most cases, these can be adaptations of existing test standards for conventional concrete but modified for the particular properties of UHPC. In addition, generic material specifications are needed to encourage the introduction of competitive materials.

Production Issues

At the present time, very few producers have experience with UHPC for precast or cast-in-place applications. Information needs to be made available so that they are aware of the differences to expect with UHPC. For example, precasters need to be aware of the need for longer mixing times in conventional concrete mixers, longer set times, and modified curing regimes. Quality control tolerances need to be defined for the standard test methods. For example, the use of small-size cylinders for measurement of compressive strength needs to be established, along with the requirements for specimen preparation and testing machine capabilities.

Research Needs

The largest general design topic area where research is lacking concerns reinforcement details for nonprestressed reinforcement and prestressing strands. This includes development of and splice lengths of bars in tension and compression. Although the existing provisions for conventional concrete could be used, they do not take advantage of the enhanced tensile and compressive strengths of UHPC. A systematic investigation of strand spacing and strand cover is needed for 0.5-, 0.6-, and 0.7-inch (12.7-, 15.2-, and 17.8-mm)-diameter strands to determine whether decreased spacing can be used with UHPC.

Investigations into the use of and reliance upon fiber reinforcement in structural concrete members is also needed. Fiber type, geometry, volume, dispersion, and orientation can all affect the structural performance of the concrete member. Development of interrelated material proportioning methods, component fabrication methods, and structural design concepts are recommended.

U.S. Federal law requires compliance with Buy America provisions. Research is needed into the use of either domestically produced steel fibers and/or the use of non-steel fibers while still producing a UHPC-class material that affords appropriate characteristics.

Summary of Needs

To encourage greater implementation of UHPC in the highway infrastructure, the following activities and documents are needed in approximate order of priority:

- Studies showing the cost effectiveness of UHPC in various applications.
- Design and construction guide specifications for structures made with UHPC.
- Research to address some of the missing information needed in the structural design guidelines.
- Standard test methods and material specifications for UHPC.
- Production procedures for precast and cast-in-place construction.
- Broader geographic distribution of demonstration projects.
- Ongoing and greater distribution of technical information.

AASHTO and FHWA should consider the development of structural design and construction guidelines. This effort should include research to address some of the needed missing information. The current efforts to engage organizations such as the American Concrete Institute, the Precast/Prestressed Concrete Institute (PCI), and ASTM should be extended to AASHTO. [309] PCI should work to develop production procedures for precast UHPC products. The National Ready Mixed Concrete Association should endeavor to address hurdles related to cast-in-place UHPC production, delivery, and casting. The involvement of the AASHTO Highway Subcommittee on Materials and ASTM Committees C09 on Concrete and Concrete Aggregates and C01 on Cement would facilitate the development of test methods and material specifications. The availability of funding to support these activities would accelerate the process.

The need for broader geographic distribution of demonstration projects should be addressed by FHWA in cooperation with the State departments of transportation.

Finally, and perhaps most important, owners need to be convinced that the use of UHPC is a good investment. Without that justification and the resulting demand, UHPC will remain a niche product.

References

[1] Graybeal, B., "Ultra-High Performance Concrete," *TechNote*, FHWA-HRT-11-038, Federal Highway Administration, McLean, VA, 2011.

[2] Blaise, P.Y. and Couture, M., "Precast, Prestressed Pedestrian Bridge—World's First Reactive Powder Concrete Structure," *PCI Journal*, Vol. 44, No. 5, September/ October 1999, pp. 60–71.

[3] Schmidt, M., "Sustainable Building With Ultra-High-Performance Concrete (UHPC)—Coordinated Research Program in Germany," *Proceedings of Hipermat 2012 3rd International Symposium on UHPC and Nanotechnology for High Performance Construction Materials*, Ed., Schmidt, M., Fehling, E., Glotzbach, C., Fröhlich, S., and Piotrowski, S., Kassel University Press, Kassel, Germany, 2012, pp. 17–25.

[4] *Bétons Fibrés à Ultra-Hautes Performances,* Recommandations Provisoires, (*Ultra High Performance Fibre-Reinforced Concretes)*, Interim Recommendations, SETRA-AFGC, Groupe de travail BFUP, Paris, France, January 2002. (in French and English). Available at http://www.afgc.asso.fr/statuts-de-lassociation-francaise-de-genie-civil.html [Cited November 23, 2011].

[5] Toutlemonde, F. et al., "Efforts et Acquis de Recherche sur les BFUP depuis 2002 en vue de l'Actualisation des Recommandations de l'AFGC," (An Overview of Research Advances From 2002 Concerning UHPFRC, in View of Updating AFGC Recommendations), *Proceedings of the International Workshop on Ultra High Performance Fibre Reinforced Concrete—Designing and Building with UHPFRC: State of the Art Development*, 17–19 November 2009, Marseille, France, AFGC/*fib*, Paper 5.3.4.

[6] Resplendino, J., "Bétons Fibrés Ultra Performant (BFUP)—Les Nouvelles Recommandations AFGC," (Updated AFGC Recommendations), *Proceedings of the International Workshop on Ultra High Performance Fibre Reinforced Concrete—Designing and Building with UHPFRC: State of the Art Development*, 17–19 November 2009, Marseille, France, AFGC/*fib*, Paper 6.1.1.

[7] Marchand, P., "Béton Fibrés Ultra Performant (BFUP) Les Nouvelles Recommandations AFGC Chapitre 2, Calcul," (Updated AFGC Recommendations: Section 2, Structural Design), *Proceedings of the International Workshop on Ultra High Performance Fibre Reinforced Concrete—Designing and Building with UHPFRC: State of the Art Development*, 17–19 November 2009, Marseille, France, AFGC/*fib*, Paper 6.1.3.

[8] Japan Society of Civil Engineers (JSCE), *Recommendations for Design and Construction of Ultra High Strength Fiber Reinforced Concrete Structures (Draft)*, JSCE Guidelines for Concrete No. 9, 2006.

[9] Kim, B.-S. et al., "R&D Activities and Application of Ultra High Performance Concrete to Cable-Stayed Bridges," *Proceedings of Hipermat 2012 3rd International Symposium on UHPC and Nanotechnology for High Performance Construction Materials*, Ed., Schmidt, M., Fehling, E., Glotzbach, C., Fröhlich, S., and Piotrowski, S., Kassel University Press, Kassel, Germany, 2012, pp. 865–872.

[10] Naaman, A.E. and Wille, K., "The Path to Ultra-High Performance Fiber Reinforced Concrete (UHP-FRC): Five Decades of Progress," *Proceedings of Hipermat 2012 3rd International Symposium on UHPC and Nanotechnology for High Performance Construction Materials*, Ed., Schmidt, M., Fehling, E., Glotzbach, C., Fröhlich, S., and Piotrowski, S., Kassel University Press, Kassel, Germany, 2012, pp. 3–15.

[11] Buitelaar, P., "Heavy Reinforced Ultra High Performance Concrete," *Proceedings of the International Symposium on Ultra High Performance Concrete*, Ed., Schmidt, M.,

Fehling, E., and GeisenhanslLike, C., Kassel University Press, Kassel, Germany, 2004, pp. 25–35.

[12] Richard, P., "Reactive Powder Concrete: A New-Ultra High Strength Cementitious Material," *Proceedings of the Fourth International Symposium on the Utilization of HighStrength/High-Performance Concrete*, 29–31 May 1996, Paris, France, Ed., de Larrard, F. and Lacroix, R., Vol. 3, pp. 1,343–1,357.

[13] Rossi, P., "Ultra-High-Performance Fiber-Reinforced Concretes. A French Perspective on Approaches Used to Produce High-Strength, Ductile Fiber-Reinforced Concrete," *Concrete International*, Vol. 23, No. 12, December 2001, pp. 46–52.

[14] Rossi, P., "Ultra High-Performance Concrete," *Concrete International*, Vol. 30, No. 2, February 2008, pp. 31–34.

[15] Racky, P., "Cost-Effectiveness and Sustainability of UHPC," *Proceedings of the International Symposium on Ultra High Performance Concrete*, Ed., Schmidt, M., Fehling, E., and GeisenhanslLike, C., Kassel University Press, Kassel, Germany, 2004, pp. 797–805.

[16] Schmidt, M. and Jerebic D., "UHPC: Basis for Sustainable Structures—The Gaertnerplatz Bridge in Kassel," *Proceedings of the Second International Symposium on Ultra High Performance Concrete*, Ed., Fehling, E., Schmidt, M., and StLirwald. S., Kassel University Press, Kassel, Germany, 2008, pp. 619–625.

[17] Sedran, T., Durand, C., and de Larrard, F., "An Example of UHPFRC Recycling," *Proceedings of the International Workshop on Ultra High Performance Fibre Reinforced Concrete—Designing and Building with UHPFRC: State of the Art Development*, 17–19 November 2009, Marseille, France, AFGC/*fib*, Paper 6.2.2.

[18] Stengel, T. and Schießl, P., "Sustainable Concrete With UHPC—From Life Cycle Inventory Data Collection to Environmental Impact Assessment," *Proceedings of the Second International Symposium on Ultra High Performance Concrete*, Ed., Fehling, E., Schmidt, M., and StLirwald, S., Kassel University Press, Kassel, Germany, 2008, pp. 461–468.

[19] Stengel, T., "Sustainability Aspects of Traffic Bridges Made From UHPFRC—State-of-theArt and Challenges for Concrete Technology," *Proceedings of the International Workshop on Ultra High Performance Fibre Reinforced Concrete—Designing and Building with UHPFRC: State of the Art Development*, 17–19 November 2009, Marseille, France, AFGC/*fib*, Paper 6.2.4.

[20] *Environmental Management Systems—General Guidelines on Principles, Systems and Support Techniques,* ISO 14040, International Organization for Standardization, Geneva, Switzerland, 2006.

[21] Piotrowski, S. and Schmidt, M., "Life Cycle Cost Analysis of a UHPC-Bridge on Example of Two Bridge Refurbishment Designs," *Proceedings of Hipermat 2012 3rd International Symposium on UHPC and Nanotechnology for High Performance Construction Materials*, Ed., Schmidt, M., Fehling, E., Glotzbach, C., Fröhlich, S., and Piotrowski, S., Kassel University Press, Kassel, Germany, 2012, pp. 957–964.

[22] Graybeal, B., "Material Property Characterization of Ultra-High Performance Concrete," FHWA, U.S. Department of Transportation, Report No. FHWA-HRT-06-103, McLean, VA, 2006.

[23] Aarup, B., "CRC—A Special Fibre Reinforced High Performance Concrete," *Proceedings of the First International RILEM Symposium on Advances in Concrete*

Through Science and Engineering, Publication pro048, March 21–24, 2004, Evanston, IL, Ed., Weiss, J. et al.

[24] Wille, K., Naaman, A.E., and Parra-Montesinos, G.J., "Ultra-High Performance Concrete With Compressive Strength Exceeding 150 MPa (22 ksi): A Simpler Way," *ACI Materials Journal*, Vol. 108, No. 1, January–February 2011, pp. 46–54.

[25] Wille, K., Naaman, A.E., and El-Tawil, S., "Optimizing Ultra-High-Performance Fiber-Reinforced Concrete," *Concrete International,* Vol. 33, No. 9, September 2011, pp. 35–41.

[26] Habel, K. et al., "Ultra-High Performance Fibre Reinforced Concrete Mix Design in Central Canada," *Canadian Journal of Civil Engineering*, Vol. 35, No. 2, February 2008, pp. 217–224.

[27] Kazemi, S. and Lubell, A., "Influence of Specimen Size and Fiber Content on Mechanical Properties of Ultra-High-Performance Fiber-Reinforced Concrete," *ACI Materials Journal*, Vol. 109, No. 6, November–December 2012, pp. 675–684.

[28] Holschemacher, K. and Weiße, D., "Economic Mix Design Ultra High-Strength Concrete," *Seventh International Symposium on the Utilization of High-Strength/High-Performance Concrete*, Vol. II, Publication No. SP-228, Ed., Russell, H.G., American Concrete Institute, Farmington Hills, MI, 2005, pp. 1,133–1,144.

[29] Plank, J., Schröfl, C., and Gruber, M., "Use of a Supplemental Agent to Improve Flowability of Ultra-High-Performance Concrete," *Ninth ACI International Conference on Superplasticizers and Other Chemical Admixtures*, Publication No. SP-262, Ed., Gupta, P., Holland, T.C., and Malhotra, V.M., American Concrete Institute, Farmington Hills, MI, 2009, pp. 1–16.

[30] Rougeau, P. and Borys, B., "Ultra High Performance Concrete With Ultrafine Particles Other Than Silica Fume," *Proceedings of the International Symposium on Ultra High Performance Concrete*, Ed., Schmidt, M., Fehling, E., and GeisenhanslLike, C., Kassel University Press, Kassel, Germany, 2004, pp. 213–225.

[31] Van, V.T.A. and Ludwig, H-M., "Proportioning Optimization of UHPC Containing Rice Husk Ash and Ground Granulated Blast-Furnace Slag," *Proceedings of Hipermat 2012 3rd International Symposium on UHPC and Nanotechnology for High Performance Construction Materials*, Ed., Schmidt, M., Fehling, E., Glotzbach, C., Fröhlich, S., and Piotrowski, S., Kassel University Press, Kassel, Germany, 2012, pp. 197–205.

[32] Arafa, M., Shihada, S., and Karmont, M., "Mechanical Properties of Ultra High Performances Concrete Produced in the Gaza Strip," *Asian Journal of Materials Science,* Vol. 2, No. 1, 2010, pp. 1–12.

[33] Fidjestol, P., Thorsteinsen, R.J., and Svennevig, P., "Making UHPC with Local Materials— The Way Forward," *Proceedings of Hipermat 2012 3rd International Symposium on UHPC and Nanotechnology for High Performance Construction Materials*, Ed., Schmidt, M., Fehling, E., Glotzbach, C., Fröhlich, S., and Piotrowski, S., Kassel University Press, Kassel, Germany, 2012, pp. 207–214.

[34] Schmidt, M. et al., "Mix Design and Properties of Ultra-High Performance Fibre Reinforced Concrete for the Construction of a Composite Concrete UHPFRC-Concrete Bridge," *Proceedings of IABSE Symposium Improving Infrastructure Worldwide—Bringing People Closer*, IABSE, Weimar, Germany, Vol. 93, 2007, pp. 466–476. Also *Ultra High Performance Concrete (UHPC)—10 Years of Research and Development at*

the University of Kassel, Ed., Schmidt, M. and Fehling, E., Kassel University Press, Kassel, Germany, 2007, pp. 44–54.

[35] Collepardi, S. et al., "Mechanical Properties of Modified Reactive Powder Concrete," *Proceedings of the Fifth CANMET/ACI International Conference on Superplasticizers and Other Chemical Admixtures in Concrete*, 1997, Rome, Italy, Publication No. SP-173, Ed., Malhotra V.M., American Concrete Institute, Farmington Hills, MI, pp. 1–21.

[36] Coppola, L. et al., "Influence of Super-Plasticizer Type on Compressive Strength of Reactive Powder Mortars," *Proceedings of the Fifth CANMET/ACI International Conference on Superplasticizers and Other Chemical Admixtures in Concrete*, 1997, Rome, Italy, Publication No. SP-173, Ed., Malhotra V.M., American Concrete Institute, Farmington Hills, MI, pp. 537–557.

[37] Teichmann, T. and Schmidt, M., "Mix Design and Durability of Ultra High Performance Concrete (UHPC)," *Proceedings of the 4th International Ph.D. Symposium in Civil Engineering*, Munich, 2002, pp. 341–347.

[38] Williams, E. et al., "Laboratory Characterization of Cor-Tuf Concrete With and Without Steel Fibers," Technical Report No. ERDC/GSL TR-02-22, U.S. Army Corps of Engineers, Engineer Research and Development Center, Washington, DC, July 2009.

[39] Roth, M.J. et al., "Laboratory Investigation of the Characterization of Cor-Tuf Flexural and Splitting Tensile Properties," Report No. ERDC/GSL TR-10-46, U.S. Army Corps of Engineers, Engineer Research and Development Center, Washington, DC, October 2010.

[40] Rossi, P. et al., "Bending and Compressive Behaviors of a New Cement Composite," *Cement and Concrete Research*, Vol. 35, No. 1, 2005, pp. 27–33.

[41] Mazanec, O. and Schieβl, P., "Mixing Time Optimization for UHPC," *Proceedings of the Second International Symposium on Ultra High Performance Concrete*, Ed., Fehling, E., Schmidt, M., and Stürwald. S., Kassel University Press, Kassel, Germany, 2008, pp. 401–408.

[42] Mazanec, O., Lowke, D., and Schieβl, P., "Mixing of High Performance Concrete: Effect of Concrete Composition and Mixing Intensity on Mixing Time," *Materials and Structure*, Vol. 43, No. 3, 2010, pp. 357–365.

[43] Kim, S.W. et al., "Effect of Filling Method on Fibre Orientation & Dispersion and Mechanical Properties of UHPC," *Proceedings of the Second International Symposium on Ultra High Performance Concrete*, Ed., Fehling, E., Schmidt, M., and Stürwald. S., Kassel University Press, Kassel, Germany, 2008, pp. 185–192.

[44] Stiel, T., Karihaloo, B., and Fehling, E., "Effect of Casting Direction on the Mechanical Properties of CARDIFRC®," *Proceedings of the International Symposium on Ultra High Performance Concrete*, Ed., Schmidt, M., Fehling, E., and Geisenhanslüke, C., Kassel University Press, Kassel, Germany, 2004, pp. 481–493.

[45] Ozyildirim, C., "Evaluation of Ultra-High-Performance Fiber-Reinforced Concrete," Virginia Center for Transportation Innovation and Research, Report No. FHWA/VCTIR 12-R1, Federal Highway Administration, McLean, VA, 2011.

[46] Graybeal, B., "Construction of Field-Cast Ultra-High Performance Concrete Connections," *TechNote*, Federal Highway Administration, McLean, VA, FHWA-HRT-12-038, 2012.

[47] Graybeal, B. and Stone, B., "Compression Response of a Rapid-Strengthening Ultra-High Performance Concrete Formulation," FHWA, U.S. Department of Transportation,

Report No. FHWA-HRT-12-065, National Technical Information Service Accession No. PB2012-112545, 2012.

[48] Ay, L., "Curing Tests on Ultra High Strength Plain and Steel Fibrous Cement Based Composites," *Proceedings of the International Symposium on Ultra High Performance Concrete*, Ed., Schmidt, M., Fehling, E., and Geisenhanslüke, C., Kassel University Press, Kassel, Germany, 2004, pp. 695–701.

[49] Heinz, D. and Ludwig, H.-M., "Heat Treatment and the Risk of DEF Delayed Ettringite Formation in UHPC," *Proceedings of the International Symposium on Ultra High Performance Concrete*, Ed., Schmidt, M., Fehling, E., and Geisenhanslüke, C., Kassel University Press, Kassel, Germany, 2004, pp. 717–730.

[50] Schachinger, I., Hilbig, H., and Stengal, T., "Effect of Curing Temperatures at an Early Age on the Long-Term Strength Development of UHPC," *Proceedings of the Second International Symposium on Ultra High Performance Concrete*, Ed., Fehling, E., Schmidt, M., and Stürwald. S., Kassel University Press, Kassel, Germany, 2008, pp. 205–212.

[51] Heinz, D., Urbonas, L., and Gerlicher, T., "Effect of Heat Treatment Method on the Properties of UHPC," *Proceedings of Hipermat 2012 3rd International Symposium on UHPC and Nanotechnology for High Performance Construction Materials*, Ed., Schmidt, M., Fehling, E., Glotzbach, C., Fröhlich, S., and Piotrowski, S., Kassel University Press, Kassel, Germany, 2012, pp. 283–290.

[52] Massidda, L. et al., "High Pressure Steam Curing of Reactive-Powder Mortars," *Fifth CANMET/ACI Conference on Recent Advances in Concrete Technology, Fifth International Conference*, Publication No. SP-200, Ed., Malhotra, V.M., American Concrete Institute, Farmington Hills, MI, 2001, pp. 447–463.

[53] Standard Test Method for Flow of Hydraulic Cement Mortar, ASTM C1437, ASTM International, Volume 04.01, West Conshohocken, PA, 2007.

[54] Standard Test Method for Slump of Hydraulic-Cement Concrete, ASTM C143, ASTM International, Volume 04.02, West Conshohocken, PA, 2012.

[55] Scheffler, B. and Schmidt, M., "Application of UHPC for Multifunctional Road Pavements," *Proceedings of Hipermat 2012 3rd International Symposium on UHPC and Nanotechnology for High Performance Construction Materials*, Ed., Schmidt, M., Fehling, E., Glotzbach, C., Fröhlich, S., and Piotrowski, S., Kassel University Press, Kassel, Germany, 2012, pp. 913–920.

[56] Swenty, M. and Graybeal, B., "Material Characterization of Field-Cast Connection Grouts," FHWA, U.S. Department of Transportation, Report No. FHWA-HRT-13-041, National Technical Information Service Accession No. PB2013-130231, 2013.

[57] Standard Method of Test for Time of Setting of Concrete Mixtures by Penetration Resistance, AASHTO Designation: T 197, *Standard Specifications for Transportation Materials and Methods of Sampling and Testing*, Part 2A—Tests, American Association of State Highway and Transportation Officials, Washington, DC, 2005.

[58] Standard Test Method for Compressive Strength of Cylindrical Concrete Specimens, ASTM C39, ASTM International, Volume 04.02, West Conshohocken, PA, 2012.

[59] Standard Practice for Capping Cylindrical Concrete Specimens, ASTM C617, ASTM International, Volume 04.02, West Conshohocken, PA, 2012.

[60] Graybeal, B. and Davis, M., "Cylinder or Cube: Strength Testing of 80 to 200 MPa (11.6 to 29 ksi) Ultra-High-Performance Fiber-Reinforced Concrete," *ACI Materials Journal*, Vol. 105, No. 6, November–December 2008, pp. 603–609.

[61] Standard Test Method for Compressive Strength of Hydraulic Cement Mortars (Using 2-in. or [50-mm] Cube Specimens), ASTM C109, ASTM International, Volume 04.02, West Conshohocken, PA, 2012.

[62] Ahlborn, T. and Kollmorgen, G., "Impact of Age and Size on the Mechanical Behavior of Ductal®," Report No. CSD-2004-07, Michigan Technological University, 2004.

[63] Graybeal, B., "UHPC for PBES Connections." Available at http://www.abc.fiu.edu/wp-content/uploads/2011/10/Graybeal-UHPC-20110929Webinar-distributed.pdf [Cited March 12, 2012].

[64] Iowa Department of Transportation, "Special Provisions for Ultra High Performance Concrete," SP-090112a, Effective Date, February 15, 2011, Available at http://www.iowadot.gov/us6kegcreek/documents/SP-concrete.pdf [Cited May 29, 2012].

[65] New York State Department of Transportation, "557.21.16—Field Cast Joints Between Precast Concrete Units," July 2010.

[66] Fröhlich, S. and Schmidt, M., "Influences on Repeatability and Reproducibility of Testing Methods for Fresh UHPC," *Proceedings of Hipermat 2012 3rd International Symposium on UHPC and Nanotechnology for High Performance Construction Materials*, Ed., Schmidt, M., Fehling, E., Glotzbach, C., Fröhlich, S., and Piotrowski, S., Kassel University Press, Kassel, Germany, 2012, pp. 225–232.

[67] Standard Test Method for Compressive Strength of Cylindrical Concrete Specimens, ASTM C39, ASTM International, Volume 04.02, West Conshohocken, PA, 2004.

[68] Orgass, M. and Klug, Y., "Fibre Reinforced Ultra-High Strength Concretes," *Proceedings of the International Symposium on Ultra High Performance Concrete*, Ed., Schmidt, M., Fehling, E., and GeisenhanslLike, C., Kassel University Press, Kassel, Germany, 2004, pp. 637–647.

[69] Magureanu, C. et al., "Mechanical Properties and Durability of Ultra-High-Performance Concrete," *ACI Materials Journal*, Vol. 109, No. 2, March–April 2012, pp. 177–183.

[70] Skazlić, M., Serdar, M., and Bjegović D., "Influence of Test Specimens Geometry on Compressive Strength of Ultra High Performance Concrete," *Proceedings of the Second International Symposium on Ultra High Performance Concrete*, Ed., Fehling, E., Schmidt, M., and Stürwald. S., Kassel University Press, Kassel, Germany, 2008, pp. 295–301.

[71] Schmidt, M. and Fröhlich, S., "Testing of UHPC," *Proceedings of the Third International fib Congress and Exhibition Incorporating the PCI Annual Convention and National Bridge Conference*, Washington, DC, May 29–June 2, 2010, Compact Disc, Paper 237.

[72] Pimienta, P. et al., "Behaviour of UHPFRC at High Temperatures," *Proceedings of the International Workshop on Ultra High Performance Fibre Reinforced Concrete—Designing and Building with UHPFRC: State of the Art Development*, 17–19 November 2009, Marseille, France, AFGC/*fib*, Paper 5.3.2.

[73] Diederichs, U. and Mertzsch, O., "Behaviour of Ultra High Strength Concrete at High Temperatures," *Proceedings of the Second International Symposium on Ultra High*

Performance Concrete, Ed., Fehling, E., Schmidt, M., and Stürwald. S., Kassel University Press, Kassel, Germany, 2008, pp. 347–354.

[74] Curbach, M. and Speck, K., "Ultra High Performance Concrete Under Biaxial Compression," *Proceedings of the Second International Symposium on Ultra High Performance Concrete*, Ed., Fehling, E., Schmidt, M., and Stürwald. S., Kassel University Press, Kassel, Germany, 2008, pp. 477–484.

[75] Leutbecher, T. and Fehling, E., "Structural Behaviour of UHPC Under Tensile Stress and Biaxial Loading," *Proceedings of the International Symposium on Ultra High Performance Concrete*, Ed., Schmidt, M., Fehling, E., and Geisenhanslüke, C., Kassel University Press, Kassel, Germany, 2004, pp. 435–446.

[76] *AASHTO LRFD Bridge Design Specifications*, Sixth Edition, American Association of State Highway and Transportation Officials, Washington, DC, 2012.

[77] Graybeal, B., Perry, V., and Royce, M., "UHPC Ultra-High Performance Concrete," NHI Innovations Webinar, November 18, 2010. Available at https://connectdot.connectsolutions.com/n134083201011 [Cited April 3, 2012].

[78] Graybeal, B. et al., "Direct and Flexural Test Methods for Determination of the Tensile Stress-Strain Response of UHPFRC," *Proceedings of Hipermat 2012 3rd International Symposium on UHPC and Nanotechnology for High Performance Construction Materials*, Ed., Schmidt, M., Fehling, E., Glotzbach, C., Fröhlich, S., and Piotrowski, S., Kassel University Press, Kassel, Germany, 2012, pp. 395–402.

[79] Graybeal, B. and Baby, F., "Development of a Direct Tension Test Method for UHPFRC," *ACI Materials Journal*, Vol. 110, No. 2, March–April 2013.

[80] Standard Test Method for Flexural Strength of Concrete (Using Simple Beam with Third-Point Loading), ASTM C78, ASTM International, Volume 04.02, West Conshohocken, PA, 2010.

[81] Standard Test Method for Splitting Tensile Strength of Cylindrical Concrete Specimens, ASTM C496, ASTM International, Volume 04.02, West Conshohocken, PA, 2011.

[82] Graybeal, B.A, "Practical Means for Determination of the Tensile Behavior of Ultra-High Performance Concrete," *Journal of ASTM International*, Vol. 3, No. 8, September 2006.

[83] Standard Test Method for Flexural Toughness and First-Crack Strength of Fiber-Reinforced Concrete (Using Beam With Third Point Loading) (Withdrawn 2006), ASTM C1018, ASTM International, Volume 04.02, West Conshohocken, PA, 1997.

[84] Standard Test Method for Flexural Performance of Fiber-Reinforced Concrete (Using Beam With Third-Point Loading), ASTM C1609, ASTM International, Volume 04.02, West Conshohocken, PA, 2012.

[85] RILEM TC 162-TDF, "Test and Design Methods for Steel Fibre Reinforced Concrete, Recommendations," *Materials and Structures*, Vol. 35, 2002, pp. 579–582.

[86] Baby, F. et al., "A Proposed Flexural Test Method and Associated Inverse Analysis for UHPFRC," *ACI Materials Journal*, Vol. 109, No. 5, September–October 2012, pp. 545–555.

[87] Qian, S. and Li, V.C., "Simplified Inverse Method for Determining the Tensile Properties of SHCCs," *Journal of Advanced Concrete Technology*, Vol. 6, No. 2, 2008, pp. 353–363.

[88] Wille, K. and Parra-Montesinos, G., "Effect of Beam Size, Casting Method, and Support Conditions on Flexural Behavior of Ultra-High-Performance Fiber-Reinforced Concrete," *ACI Materials Journal*, Vol. 109, No. 3, May–June 2012, pp. 379–388.

[89] Standard Method of Test for Tensile Strength of Hydraulic Cement Mortars, AASHTO Designation: T 132, *Standard Specifications for Transportation Materials and Methods of Sampling and Testing*, Part 2A—Tests, American Association of State Highway and Transportation Officials, Washington, DC, 1987.

[90] Reineck, K.-H. and Frettlöhr, B., "Tests on Scale Effect of UHPFRC Under Bending and Axial Forces," *Proceedings of the Third International fib Congress and Exhibition Incorporating the PCI Annual Convention and National Bridge Conference*, Washington, DC, May 29–June 2, 2010, Compact Disc, Paper 54.

[91] Gao, R., Stroeven, P., and Hendriks, C.F., "Mechanical Properties of Reactive Powder Concrete Beams," *Seventh International Symposium on the Utilization of High-Strength/High-Performance Concrete*, Vol. II, Publication No. SP-228, Ed., Russell, H.G., American Concrete Institute, Farmington Hills, MI, 2005, pp. 1,237–1,252.

[92] Ichinomiya, T. et al., "Experimental Study on Mechanical Properties of Ultra-High-Strength Concrete with Low-Autogenous-Shrinkage," *Seventh International Symposium on the Utilization of High-Strength/High-Performance Concrete*, Vol. II, Publication No. SP-228, Ed., Russell, H.G., American Concrete Institute, Farmington Hills, MI, 2005, pp. 1,341–1,352.

[93] Lallemant-Gamboa, I. et al., "Formulations, Characterizations and Applications of Ultra High-Performance Concrete," *Seventh International Symposium on the Utilization of HighStrength/High-Performance Concrete*, Vol. II, Publication No. SP-228, Ed., Russell, H.G., American Concrete Institute, Farmington Hills, MI, 2005, pp. 1,221–1,235.

[94] Marijan, S., Dubravka, B., and Zeljana, S., "Acoustic Emission Response and Mechanical Characterization of Ultra High-Performance Concrete Types," *Seventh International Symposium on the Utilization of High-Strength/High-Performance Concrete*, Vol. II, Publication No. SP-228, Ed., Russell, H.G., American Concrete Institute, Farmington Hills, MI, 2005, pp. 1,207–1,220.

[95] Leutbecher, T. and Fehling E., "Tensile Behavior of Ultra-High-Performance Concrete Reinforced with Reinforcing Bars and Fibers: Minimizing Fiber Content," *ACI Materials Journal*, Vol. 109, No. 2, March–April 2012, pp. 253–263.

[96] Roth, M.J. et al., "Ultra-High Strength, Glass Fiber-Reinforced Concrete: Mechanical Behavior and Numerical Modeling," *ACI Materials Journal*, Vol. 107, No. 2, March–April 2010, pp. 185–194.

[97] Zheng, W., Kwan, A.K.H., and Lee, P.K.K., "Direct Tension Test of Concrete," *ACI Materials Journal*, Vol. 98, No. 1, January–February 2001, pp. 63–71.

[98] Zhang, J., Stang, H., and Li, V., "Experimental Study on Crack Bridging in FRC Under Uniaxial Fatigue Tension," *Journal of Materials in Civil Engineering*, ASCE, Vol. 12, No. 1, February 2000, pp. 66–73.

[99] Nielsen, C.V., "Tensile Postcrack Behavior of Steel Fiber Reinforced Ultra-High Strength Concrete," *International Workshop on High Performance Concrete,* Publication No. SP-159, Ed., Zia, P., American Concrete Institute, Farmington Hills, MI, 1996, pp. 231–246.

[100] Behloul, M., Bernier, G., and Cheyrezy, M., "Tensile Behavior of Reactive Powder Concrete (RPC)," *Proceedings of the Fourth International Symposium on the Utilization of HighStrength/High-Performance Concrete*, 29–31 May 1996, Paris, France, Ed. de Larrard, F. and Lacroix, R., Vol. 3, pp. 1,375–1,381.

[101] Wang, Y., Li, V.C., and Backer, S., "Experimental Determination of Tensile Behavior of Fiber Reinforced Concrete," *ACI Materials Journal*, Vol. 87, No. 5, September–October 1990, pp. 461–468.

[102] D'Alessandro, K.C. et al., "Investigation of Biaxial Stress States of Ultra-High Performance Concrete Bridge Girders Through Small Panel Testing and Finite Element Analysis," *Proceedings of the International Workshop on Ultra High Performance Fibre Reinforced Concrete—Designing and Building with UHPFRC: State of the Art and Development*, 17-18 November 2009, Marseille, France, AFGC/*fib*, Paper P 2.

[103] Standard Test Method for Static Modulus of Elasticity and Poisson's Ratio of Concrete in Compression, ASTM C469, ASTM International, Volume 04.02, West Conshohocken, PA, 2010.

[104] Ma, J. et al., "Comparative Investigations on Ultra-High Performance Concrete With and Without Coarse Aggregates," *Proceedings of the International Symposium on Ultra High Performance Concrete*, Ed., Schmidt, M., Fehling, E., and GeisenhanslLike, C., Kassel University Press, Kassel, Germany, 2004, pp. 205–212.

[105] Bonneau, O. et al., "Mechanical Properties and Durability of Two Industrial Reactive Powder Concretes," *ACI Materials Journal*, Vol. 94, No. 4, July–August 1997, pp. 286–290.

[106] Simon, A., "Les Nouvelles Recommandations AFGC sur les BFUP CHAPITRE I—Comportement et Caractéristiques Mécaniques des BFUP," (Updated AFGC Recommendations: Section 1 Materials), *Proceedings of the International Workshop on Ultra High Performance Fibre Reinforced Concrete—Designing and Building with UHPFRC: State of the Art Development*, 17–19 November 2009, Marseille, France, AFGC/*fib*, Paper 6.1.2.

[107] Joh, C. et al., "Punching Shear Strength Estimation of UHPC Slabs," *Proceedings of the Second International Symposium on Ultra High Performance Concrete*, Ed., Fehling, E., Schmidt, M., and StLirwald. S., Kassel University Press, Kassel, Germany, 2008, pp. 719-726.

[108] Ahlborn, T.M. et al., "Durability and Strength Characterization of Ultra-High Performance Concrete Under Variable Curing Regimes," *Proceedings of the Second International Symposium on Ultra High Performance Concrete*, Ed., Fehling, E., Schmidt, M., and StLirwald, S., Kassel University Press, Kassel, Germany, 2008, pp. 197–204.

[109] Graybeal, B., "Structural Behavior of Ultra-High Performance Concrete Prestressed I-Girders," FHWA, U.S. Department of Transportation, Report No. FHWA-HRT-06-115, 2006. Available through National Technical Information Service at http://www.ntis.gov with Accession No. PB2007104386 [Cited November 23, 2011].

[110] Ocel, J. and Graybeal, B., "Fatigue Behavior of an Ultra-High Performance Concrete I-Girder," *Proceedings of the PCI National Bridge Conference*, October 22–24, 2007, Phoenix, AZ, Compact Disc, Paper 82.

[111] Parsekian, G.A. et al., "Static and Fatigue Tests on Ductal® UHPFRC Footbridge Sections," *Proceedings of the Fifth ACI/CANMET/IBRACON International Conference*

on High-Performance Concrete Structures and Materials, 18–20 June 2008, Manaus, Amazon State (AM), Brazil, Publication No. SP-253, Ed., Figueiredo, E.P. et al., American Concrete Institute, Farmington Hills, MI, 2008, pp. 273–290.

[112] Bierwagen, D. et al., "Ultra-High Performance Concrete Waffle Slab Bridge Deck for Wapello County, Iowa," *HPC Bridge Views*, Issue No. 65, January/February 2011. Available at http://www.hpcbridgeviews.org [Cited November 23, 2011].

[113] Aaleti, S. et al., "Experimental Evaluation of Structural Behavior of Precast UHPC Waffle Bridge Deck Panels and Connections," *Transportation Research Board 90th Annual Meeting Compendium of Papers*, January 23–27, 2011, Washington, DC, Compact Disc, Session 404, Paper 11-2705.

[114] Graybeal, B. and Hartmann, J., "Ultra-High Performance Concrete Material Properties," Transportation Research Board Annual Meeting, 2003, Washington, DC, Compact Disc.

[115] Schmidt, M. et al., "Durability of Ultra High Performance Concrete," *Proceedings of the 6th International Symposium on High Strength/High Performance Concrete*, Leipzig, Germany, June 2002, Ed., König, G., Dehn, F., and Faust, T., Vol. 2, pp. 1,367-1,376.

[116] Fitik, B., Niedermeier, R., and Zilch, K., "Fatigue Behaviour of Ultra High-Performance Concrete Under Cyclic Stress Reversal Loading," *Proceedings of the Second International Symposium on Ultra High Performance Concrete*, Ed., Fehling, E., Schmidt, M., and Stürwald. S., Kassel University Press, Kassel, Germany, 2008, pp. 529–536.

[117] Grünberg, J. et al., "Multi-Axial and Fatigue Behaviour of Ultra-High-Performance Concrete," *Proceedings of the Second International Symposium on Ultra High Performance Concrete*, Ed., Fehling, E., Schmidt, M., and Stürwald. S., Kassel University Press, Kassel, Germany, 2008, pp. 485–492.

[118] Lohaus, L. and Elsmeier, K., "Fatigue Behaviour of Plain and Fibre Reinforced Ultra-High Performance Concrete," *Proceedings of Hipermat 2012 3rd International Symposium on UHPC and Nanotechnology for High Performance Construction Materials*, Ed., Schmidt, M., Fehling, E., Glotzbach, C., Fröhlich, S., and Piotrowski, S., Kassel University Press, Kassel, Germany, 2012, pp. 631–637.

[119] Behloul, M. et al., "Fatigue Flexural Behavior of Pre-Cracked Specimens of Special UHPFRC," *Seventh International Symposium on the Utilization of High-Strength/High-Performance Concrete*, Vol. II, Publication No. SP-228, Ed., Russell, H.G., American Concrete Institute, Farmington Hills, MI, 2005, pp.1,253–1,268.

[120] Lappa, E.S., Braam, C.R., and Walraven, J.C., "Static and Fatigue Bending Tests of UHPC," *Proceedings of the International Symposium on Ultra High Performance Concrete*, Ed., Schmidt, M., Fehling, E., and GeisenhanslLike, C., Kassel University Press, Kassel, Germany, 2004, pp. 449–458.

[121] Fehling, E., Bunje, K., and Leutbecher, T., "Design Relevant Properties of Hardened Ultra High Performance Concrete," *Proceedings of the International Symposium on Ultra High Performance Concrete*, Ed., Schmidt, M., Fehling, E., and GeisenhanslLike, C., Kassel University Press, Kassel, Germany, 2004, pp. 327–338.

[122] Behloul, M. et al., "Fire Resistance of Ductal® Ultra High Performance Concrete," *Proceedings of the First fib Congress*, Osaka, Japan, 2002, pp. 421–430.

[123] Carbonell, M.A. et al., "Bond Strength Between UHPC and Normal Strength Concrete (NSC) in Accordance with Split Prism and Freeze-Thaw Cycling Tests," *Proceedings of Hipermat 2012 3rd International Symposium on UHPC and Nanotechnology for High Performance Construction Materials*, Ed., Schmidt, M., Fehling, E., Glotzbach, C., Fröhlich, S., and Piotrowski, S., Kassel University Press, Kassel, Germany, 2012, pp. 377–384.

[124] Standard Test Method for Resistance of Concrete to Rapid Freezing and Thawing, ASTM C666, ASTM International, Volume 04.02, West Conshohocken, PA, 2008.

[125] Bindiganvile, V., Bantha, N., and Aarup, B., "Impact Response of Ultra-High-Strength Fiber-Reinforced Cement Composite," *ACI Materials Journal*, Vol. 99, No. 6, November–December 2002, pp. 543–548.

[126] Cadoni, E., Caverzan, A., and di Prisco, M., "Dynamic Behaviour of HPFR Cementitious Composites," *Proceedings of the Second International Symposium on Ultra High Performance Concrete*, Ed., Fehling, E., Schmidt, M., and StLirwald. S., Kassel University Press, Kassel, Germany, 2008, pp. 743–750.

[127] Leonhardt, S., Lowke, D., and Gehlen, G., "Effect of Fibres on Impact Resistance of Ultra High Performance Concrete," *Proceedings of Hipermat 2012 3rd International Symposium on UHPC and Nanotechnology for High Performance Construction Materials*, Ed., Schmidt, M., Fehling, E., Glotzbach, C., Fröhlich, S., and Piotrowski, S., Kassel University Press, Kassel, Germany, 2012, pp. 811–817.

[128] Standard Test Method for Creep of Concrete in Compression, ASTM C512, ASTM International, Volume 04.02, West Conshohocken, PA, 2010.

[129] Burkart, I. and MLiller, H.S., "Creep and Shrinkage Characteristics of Ultra High Strength Concrete (UHPC)," *Proceedings of the Second International Symposium on Ultra High Performance Concrete*, Ed., Fehling, E., Schmidt, M., and StLirwald. S., Kassel University Press, Kassel, Germany, 2008, pp. 469–476.

[130] Acker, P. and Behloul, M., "Ductal® Technology: A Large Spectrum of Properties, A Wide Range of Applications," *Proceedings of the International Symposium on Ultra High Performance Concrete*, Ed., Schmidt, M., Fehling, E., and GeisenhanslOke, C., Kassel University Press, Kassel, Germany, 2004, pp. 11–23.

[131] Francisco, P. et al., "Ultra High Performance Concrete for Prestressed Elements—Interest of Creep Prediction," *Proceedings of the International Workshop on Ultra High Performance Fibre Reinforced Concrete—Designing and Building with UHPFRC: State of the Art Development*, 17–19 November 2009, Marseille, France, AFGC/*fib*, Paper 5.2.1.

[132] Francisco, P. et al., "Creep and Shrinkage Prediction for a Heat-Treated Ultra High Performance Fibre-Reinforced Concrete," *Proceedings of Hipermat 2012 3rd International Symposium on UHPC and Nanotechnology for High Performance Construction Materials*, Ed., Schmidt, M., Fehling, E., Glotzbach, C., Fröhlich, S., and Piotrowski, S., Kassel University Press, Kassel, Germany, 2012, pp. 325–331.

[133] Flietstra, J.C. et al., "Creep Behavior of UHPC Under Compressive Loading With Varying Curing Regimes," *Proceedings of Hipermat 2012 3rd International Symposium on UHPC and Nanotechnology for High Performance Construction Materials*, Ed., Schmidt, M., Fehling, E., Glotzbach, C., Fröhlich, S., and Piotrowski, S., Kassel University Press, Kassel, Germany, 2012, pp. 333–340.

[134] Standard Test Method for Length Change of Hardened Hydraulic-Cement Mortar and Concrete, ASTM C157, ASTM International, Volume 04.02, West Conshohocken, PA, 2008.

[135] Eppers, S. and MOller, C., "Autogenous Shrinkage Strain of Ultra-High-Performance Concrete (UHPC)," *Proceedings of the Second International Symposium on Ultra High Performance Concrete*, Ed., Fehling, E., Schmidt, M., and StOrwald. S., Kassel University Press, Kassel, Germany, 2008, pp. 433–441.

[136] Staquet, S. and Espion, B., "Early-Age Autogenous Shrinkage of UHPC Incorporating Very Fine Fly Ash or Metakaolin in Replacement of Silica Fume," *Proceedings of the International Symposium on Ultra High Performance Concrete*, Ed., Schmidt, M., Fehling, E., and GeisenhanslOke, C., Kassel University Press, Kassel, Germany, 2004, pp. 587–599.

[137] Suzuki, M., Maruyama, I., and Sato, R., "Properties of Expansive-Ultra High-Strength Concrete," *Seventh International Symposium on the Utilization of High-Strength/High-Performance Concrete*, Vol. II, Publication No. SP-228, Ed., Russell, H.G., American Concrete Institute, Farmington Hills, MI, 2005, pp. 1,159–1,173.

[138] Kim, S. et al., "Shrinkage Behavior of Ultra High Performance Concrete at the Manufacturing Stage," *Proceedings of Hipermat 2012 3rd International Symposium on UHPC and Nanotechnology for High Performance Construction Materials*, Ed., Schmidt, M., Fehling, E., Glotzbach, C., Fröhlich, S., and Piotrowski, S., Kassel University Press, Kassel, Germany, 2012, pp. 317–324.

[139] Graybeal, B.A., "Structural Behavior of a Prototype Ultra-High Performance Concrete Pi-Girder," FHWA, U.S. Department of Transportation, Report No. FHWA-HRT-10-027, 2009. Available through National Technical Information Service at http://www.ntis.gov with Accession No. PB2009115495 [Cited November 23, 2011].

[140] Meade, T.M. and Graybeal, B.A., "Flexural Response of Lightly Reinforced Ultra-High Performance Concrete Beams," *Proceedings of the Third International fib Congress and Exhibition Incorporating the PCI Annual Convention and National Bridge Conference*, Washington, DC, May 29–June 2, 2010, Compact Disc, Paper 101.

[141] Visage, E.T. et al., "Experimental and Analytical Analysis of the Flexural Behavior of UHPC Beams," *Proceedings of Hipermat 2012 3rd International Symposium on UHPC and Nanotechnology for High Performance Construction Materials*, Ed., Schmidt, M., Fehling, E., Glotzbach, C., Fröhlich, S., and Piotrowski, S., Kassel University Press, Kassel, Germany, 2012, pp. 403–410.

[142] Gröger, J., Viet Tue, N., and Wille, K., "Bending Behaviour and Variation of Flexural Parameters of UHPFRC," *Proceedings of Hipermat 2012 3rd International Symposium on UHPC and Nanotechnology for High Performance Construction Materials*, Ed., Schmidt, M., Fehling, E., Glotzbach, C., Fröhlich, S., and Piotrowski, S., Kassel University Press, Kassel, Germany, 2012, pp. 419–426.

[143] Frettlöhr, B. et al., "Tests on the Flexural Tensile Strength of a UHPFRC Subjected to Cycling and Reversed Loading," *Proceedings of Hipermat 2012 3rd International Symposium on UHPC and Nanotechnology for High Performance Construction Materials*, Ed., Schmidt, M., Fehling, E., Glotzbach, C., Fröhlich, S., and Piotrowski, S., Kassel University Press, Kassel, Germany, 2012, pp. 427–434.

[144] Stürwald, S. and Fehling, E., "Design of Reinforced UHPFRC in Flexure," *Proceedings of Hipermat 2012 3rd International Symposium on UHPC and Nanotechnology for*

High Performance Construction Materials, Ed., Schmidt, M., Fehling, E., Glotzbach, C., Fröhlich, S., and Piotrowski, S., Kassel University Press, Kassel, Germany, 2012, pp. 443–450.

[145] Adeline, R. and Behloul, M., "High Ductile Beams Without Passive Reinforcement," *Proceedings of the Fourth International Symposium on the Utilization of High-Strength/ High-Performance Concrete*, 29-31 May 1996, Paris, France, Ed., de Larrard, F. and Lacroix, R., Vol. 3, pp. 1,383–1,390.

[146] Maguire, M. et al., "Ultra-High-Performance Concrete in Standard Precast/Prestressed Concrete Products," *Proceedings of the PCI National Bridge Conference*, September 12–15, 2009, San Antonio, TX, Paper 33.

[147] Steinberg, E. and Reeves, E., "Structural Reliability of UHPC Bridge Girders in Flexure," *Proceedings of the PCI National Bridge Conference*, October 17–20, 2004, Atlanta, GA, Compact Disc, Paper 29.

[148] Reeves, E.E., "Structural Reliability of Ultra-High Performance Concrete in Flexure," Master's Thesis, College of Engineering and Technology, Ohio University, Athens, OH, 2004.

[149] Steinberg, E., "Structural Reliability of Prestressed UHPC Flexure Models for Bridge Girders," *Journal of Bridge Engineering*, Vol. 15, No. 1, 2010, pp. 65–72.

[150] Voo, Y.L. and Foster, S.J., "Malaysia First Ultra-High Performance Concrete Prestressed Motorway Bridge: Experimental Verification," 5th International Specialty Conference on Fiber Reinforced Materials, Singapore, August 28–29, 2008. Available at http://www.ju.edu.jo/sites/Academic/hunaiti/Lists/Published%20Research /Attach ments/6/Ab stract.pdf [Cited March 24, 2012].

[151] Sujivorakul, C., "Flexural Model of Doubly Reinforced Concrete Beams Using Ultra High Performance Fiber Reinforced Concrete," *Proceedings of Hipermat 2012 3rd International Symposium on UHPC and Nanotechnology for High Performance Construction Materials*, Ed., Schmidt, M., Fehling, E., Glotzbach, C., Fröhlich, S., and Piotrowski, S., Kassel University Press, Kassel, Germany, 2012, pp. 435–442.

[152] Walsh, K.K. and Steinberg E.P., "Moment Redistribution Capacity in Ultra-High Performance Concrete," *Proceedings of Hipermat 2012 3rd International Symposium on UHPC and Nanotechnology for High Performance Construction Materials*, Ed., Schmidt, M., Fehling, E., Glotzbach, C., Fröhlich, S., and Piotrowski, S., Kassel University Press, Kassel, Germany, 2012, pp. 639–646.

[153] Tue, N.V. et al., "Bearing Capacity of Stub Columns Made of NSC, HSC and UHPC Confined by a Steel Tube," *Proceedings of the International Symposium on Ultra High Performance Concrete*, Ed., Schmidt, M., Fehling, E., and GeisenhanslLike, C., Kassel University Press, Kassel, Germany, 2004, pp. 339–350.

[154] Empelmann, M., Teutsch, M., and Steven, G., "Load-Bearing Behaviour of Centrically Loaded UHPFRC Columns," *Proceedings of the Second International Symposium on Ultra High Performance Concrete*, Ed., Fehling, E., Schmidt, M., and StLirwald. S., Kassel University Press, Kassel, Germany, 2008, pp. 521–528.

[155] Yan, P.Y. and Feng, J.W., "Mechanical Behavior of UHPC and UHPC Filled Steel Tubular Stub Columns," *Proceedings of the Second International Symposium on Ultra High Performance Concrete*, Ed., Fehling, E., Schmidt, M., and StLirwald. S., Kassel University Press, Kassel, Germany, 2008, pp. 355–362.

[156] Jungwirth, J. and Muttoni, A., "Structural Behavior of Tension Members in UHPC," *Proceedings of the International Symposium on Ultra High Performance Concrete*, Ed., Schmidt, M., Fehling, E., and GeisenhanslLike, C., Kassel University Press, Kassel, Germany, 2004, pp. 533–544.

[157] Holschemacher, K. et al., "Ultra High Strength Concrete Under Concentrated Loading," *Proceedings of the International Symposium on Ultra High Performance Concrete*, Ed., Schmidt, M., Fehling, E., and GeisenhanslLike, C., Kassel University Press, Kassel, Germany, 2004, pp. 471–480.

[158] Holschemacher, K. et al., "Experimental Investigation on Ultra High-Strength Concrete Under Concentrated Loading," *Seventh International Symposium on the Utilization of HighStrength/High-Performance Concrete*, Vol. II, Publication No. SP-228, Ed., Russell, H.G., American Concrete Institute, Farmington Hills, MI, 2005, pp. 1,145–1,157.

[159] *Tragwerke aus Beton, Stahlbeton und Spannbeton,* DIN 1045-01, Deutsches Institut fLir Normung E.V., Berlin, Germany, 2004.

[160] Hegger, J. et al., "Connections of Precast UHPC Elements," *Proceedings of Hipermat 2012 3rd International Symposium on UHPC and Nanotechnology for High Performance Construction Materials*, Ed., Schmidt, M., Fehling, E., Glotzbach, C., Fröhlich, S., and Piotrowski, S., Kassel University Press, Kassel, Germany, 2012, pp. 655–662.

[161] Morcous, G., Maguire, M., and Tadros, M.K., "Shear Capacity of Ultra-High-Performance Concrete I-Girders With Orthogonal Welded Wire Reinforcement," *Proceedings of the Thomas T. C. Hsu Symposium: Shear and Torsion in Concrete Structures*, New Orleans, LA, Publication No. SP-265, American Concrete Institute, Farmington Hills, MI, 2009, Compact Disc, pp. 511–532.

[162] American Concrete Institute (ACI) Committee 318, Building Code Requirements for Structural Concrete (ACI 318-08) and Commentary (ACI 318R-08), ACI, Farmington Hills, MI, 2008.

[163] Baby, F. et al, "Ultimate Shear Strength of Ultra High Performance Fibre Reinforced Concrete Beams," *Proceedings of Hipermat 2012 3rd International Symposium on UHPC and Nanotechnology for High Performance Construction Materials*, Ed., Schmidt, M., Fehling, E., Glotzbach, C., Fröhlich, S., and Piotrowski, S., Kassel University Press, Kassel, Germany, 2012, pp. 485–492.

[164] Baby, F. et al., "Shear Resistance of Ultra High Performance Fibre Reinforced Concrete I-Beams," *Proceedings of the 7th International Conference on Fracture Mechanics of Concrete and Concrete Structures*, Jeju, Korea, May 2010, pp. 1,411–1,417.

[165] Bunje, K. and Fehling, E., "About Shear Force and Punching Shear Resistance of Structural Elements of Ultra High Performance Concrete," *Proceedings of the International Symposium on Ultra High Performance Concrete*, Ed., Schmidt, M., Fehling, E., and Geisenhanslüke, C., Kassel University Press, Kassel, Germany, 2004, pp. 401–411.

[166] Hegger, J., Tuchlinski, D., and Komner, B., "Bond Anchorage Behavior and Shear Capacity of Ultra High Performance Concrete Beams," *Proceedings of the International Symposium on Ultra High Performance Concrete*, Ed., Schmidt, M., Fehling, E., and Geisenhanslüke, C., Kassel University Press, Kassel, Germany, 2004, pp. 351–360.

[167] Hegger, J. and Bertram, G., "Shear Carrying Capacity of Steel Fiber Reinforced UHPC," *Proceedings of the Second International Symposium on Ultra High Performance Concrete*, Ed., Fehling, E., Schmidt, M., and Stürwald. S., Kassel University Press, Kassel, Germany, 2008, pp. 513–520.

[168] Cauberg. N. et al., "Shear Capacity of UHPC—Beam Tests," *Proceedings of Hipermat 2012 3rd International Symposium on UHPC and Nanotechnology for High Performance Construction Materials*, Ed., Schmidt, M., Fehling, E., Glotzbach, C., Fröhlich, S., and Piotrowski, S., Kassel University Press, Kassel, Germany, 2012, pp. 451–458.

[169] Fehling, E. and Thiemicke, J., "Experimental Investigation on I-Shaped UHPC Beams With Combined Reinforcement Under Shear Load," *Proceedings of Hipermat 2012 3rd International Symposium on UHPC and Nanotechnology for High Performance Construction Materials*, Ed., Schmidt, M., Fehling, E., Glotzbach, C., Fröhlich, S., and Piotrowski, S., Kassel University Press, Kassel, Germany, 2012, pp. 477–484.

[170] Bertram, G. and Hegger, J., "Shear Behavior of Pretensioned UHPC Beams—Tests and Design," *Proceedings of Hipermat 2012 3rd International Symposium on UHPC and Nanotechnology for High Performance Construction Materials*, Ed., Schmidt, M., Fehling, E., Glotzbach, C., Fröhlich, S., and Piotrowski, S., Kassel University Press, Kassel, Germany, 2012, pp. 493–500.

[171] Bertram, G. and Hegger, J., "Pretensioned UHPC Beams With and Without Openings," *Proceedings of the Third International fib Congress and Exhibition Incorporating the PCI Annual Convention and National Bridge Conference*, Washington, DC, May 29–June 2, 2010, Compact Disc, Paper 236.

[172] Wu, X. and Han, S.-M., "First Diagonal Cracking and Ultimate Shear of I-Shaped Reinforced Girders of Ultra High Performance Fiber Reinforced Concrete Without Stirrup," *International Journal of Concrete Structures and Materials*, Vol. 3, No. 1, 2009, pp. 47–56.

[173] Harris, D.K. and Roberts-Wollmann, C.L., "Characterization of Punching Shear Capacity of Thin Ultra-High Performance Concrete Slabs," *Proceedings of the Second International Symposium on Ultra High Performance Concrete*, Ed., Fehling, E., Schmidt, M., and Stürwald. S., Kassel University Press, Kassel, Germany, 2008, pp. 727–734.

[174] Harris, D.K. and Roberts-Wollmann, C.L., "Characterization of the Punching Shear Capacity of Thin Ultra-High Performance Concrete Slabs," Virginia Transportation Research Council, Report No. VTRC 05-CR26, Richmond, VA, 2005.

[175] Toutlemonde, F. et al., "Innovative Design of Ultra High-Performance Fibre Reinforced Concrete Ribbed Slab: Experimental Validation and Preliminary Detailed Analyses," *Seventh International Symposium on the Utilization of High-Strength/High-Performance Concrete*, Vol. II, Publication No. SP-228, Ed., Russell, H.G., American Concrete Institute, Farmington Hills, MI, 2005, pp. 1,187–1,205.

[176] Naaman, A., Likhitruangsilp, V., and Parra-Montesinos, G., "Punching Shear Response of High-Performance Fiber-Reinforced Cementitious Composite Slabs," *ACI Structural Journal*, Vol. 104, No. 2, March–April 2007, pp. 170–179.

[177] American Concrete Institute (ACI) Committee 318, Building Code Requirements for Structural Concrete (ACI 318-05) and Commentary (ACI 318R-05), ACI, Farmington Hills, MI, 2005.

[178] Saleem, M.A. et al., "Ultra-High Performance Concrete Light-Weight Deck for Moveable Bridges," *Proceedings of the Third International fib Congress and Exhibition Incorporating the PCI Annual Convention and National Bridge Conference*, Washington, DC, May 29–June 2, 2010, Compact Disc, Paper 361.

[179] Aaleti, S.R. et al., "Precast UHPC Waffle Deck Panels and Connections for Accelerated Bridge Construction," *Proceedings of the PCI National Bridge Conference*, October 22–26, 2011, Salt Lake City, UT, Compact Disc, Paper 84.

[180] Moreillon, L., Nseir, J., and Suter, R., "Shear and Flexural Strength of Thin UHPC Slabs," *Proceedings of Hipermat 2012 3rd International Symposium on UHPC and Nanotechnology for High Performance Construction Materials*, Ed., Schmidt, M., Fehling, E., Glotzbach, C., Fröhlich, S., and Piotrowski, S., Kassel University Press, Kassel, Germany, 2012, pp. 749–756.

[181] Banta, T.E., "Horizontal Shear Transfer Between Ultra High Performance Concrete and Lightweight Concrete," Master's Thesis, Virginia Polytechnic Institute and State University, 2005.

[182] *AASHTO LRFD Bridge Design Specifications,* Third Edition, American Association of State Highway and Transportation Officials, Washington, DC, 2004.

[183] Crane, C.K. and Kahn, L.F., "Interface Shear Capacity of Small UHPC/HPC Composite T-Beams," *Proceedings of Hipermat 2012 3rd International Symposium on UHPC and Nanotechnology for High Performance Construction Materials*, Ed., Schmidt, M., Fehling, E., Glotzbach, C., Fröhlich, S., and Piotrowski, S., Kassel University Press, Kassel, Germany, 2012, pp. 459–467.

[184] Graybeal, B., "Ultra-High Performance Concrete Composite Connections for Precast Concrete Bridge Decks," FHWA, U.S. Department of Transportation, Report No. FHWAHRT-12-041, 109 pp. Available through National Technical Information Service at http://www.ntis.gov with Accession No. PB2012107569 [Cited April 13, 2012]. To be published.

[185] Hegger, J., Rauscher, S., and Goralski, C., "Push-Out Tests on Headed Studs Embedded in UHPC," *Proceedings of the International Symposium on Ultra High Performance Concrete*, Ed., Schmidt, M., Fehling, E., and GeisenhanslLike, C., Kassel University Press, Kassel, Germany, 2004, pp. 425–434.

[186] Hegger, J. and Rauscher, S., "UHPC in Composite Construction," *Proceedings of the Second International Symposium on Ultra High Performance Concrete*, Ed., Fehling, E., Schmidt, M., and StLirwald. S., Kassel University Press, Kassel, Germany, 2008, pp. 545–552.

[187] Hegger, J., Rauscher, S., and Gallwoszus, J., "Modern Hybrid Structures Made of UHPC and High Strength Steel," *Proceedings of the Third International fib Congress and Exhibition Incorporating the PCI Annual Convention and National Bridge Conference*, Washington, DC, May 29–June 2, 2010, Compact Disc, Paper 52.

[188] Jungwirth, J. et al., "Utilization of UHPC in Composite Structures—Lightweight Composite Structures (LCS)," *Proceedings of the Second International Symposium on Ultra High Performance Concrete*, Ed., Fehling, E., Schmidt, M., and StLirwald. S., Kassel University Press, Kassel, Germany, 2008, pp. 887–894.

[189] Kohlmeyer, C. et al., "Investigations on Embedded Shear Connectors for Lightweight Composite Structures," *Proceedings of the Third International fib Congress and*

Exhibition Incorporating the PCI Annual Convention and National Bridge Conference, Washington, DC, May 29–June 2, 2010, Compact Disc, Paper 429.

[190] Fehling, E. and Ismail, M., "Experimental Investigations on UHPC Structural Elements Subject to Pure Torsion," *Proceedings of Hipermat 2012 3rd International Symposium on UHPC and Nanotechnology for High Performance Construction Materials*, Ed., Schmidt, M., Fehling, E., Glotzbach, C., Fröhlich, S., and Piotrowski, S., Kassel University Press, Kassel, Germany, 2012, pp. 501–508.

[191] Joh, C. et al., "Torsional Test of Ultra High Performance Fiber-Reinforced Concrete Square Members," *Proceedings of Hipermat 2012 3rd International Symposium on UHPC and Nanotechnology for High Performance Construction Materials*, Ed., Schmidt, M., Fehling, E., Glotzbach, C., Fröhlich, S., and Piotrowski, S., Kassel University Press, Kassel, Germany, 2012, pp. 509–516.

[192] Empelmann, M. and Oettel, V., "UHPFRC Box Girders Under Torsion," *Proceedings of Hipermat 2012 3rd International Symposium on UHPC and Nanotechnology for High Performance Construction Materials*, Ed., Schmidt, M., Fehling, E., Glotzbach, C., Fröhlich, S., and Piotrowski, S., Kassel University Press, Kassel, Germany, 2012, pp. 517–524.

[193] Graybeal, B., "Behavior of Field-Cast Ultra-High Performance Concrete Bridge Deck Connections Under Cyclic and Static Structural Loading," FHWA, U.S. Department of Transportation, Report No. FHWA-HRT-11-023, National Technical Information Service Accession No. PB2011-101995, 2010.

[194] Swenty, M. and Graybeal, B., "Influence of Differential Deflection on Staged Construction Deck-Level Connections," FHWA, U.S. Department of Transportation, Report No. FHWAHRT-12-057, National Technical Information Service Accession No. PB2012-111528, 2012.

[195] Holschemacher, K., Weiβe, D., and Klotz, S., "Bond of Reinforcement in Ultra High Strength Concrete," *Proceedings of the International Symposium on Ultra High Performance Concrete*, Ed., Schmidt, M., Fehling, E., and Geisenhanslüke, C., Kassel University Press, Kassel, Germany, 2004, pp. 375–387.

[196] Holschemacher, K., Weiße, D., and Klotz, S., "Bond of Reinforcement in Ultra High-Strength Concrete," *Seventh International Symposium on the Utilization of High-Strength/High-Performance Concrete*, Vol. I, Publication No. SP-228, Ed., Russell, H.G., American Concrete Institute, Farmington Hills, MI, 2005, pp. 513–528.

[197] Fehling, E., Lorenz, P., and Leutbecher, T., "Experimental Investigations on Anchorage of Rebars in UHPC," *Proceedings of Hipermat 2012 3rd International Symposium on UHPC and Nanotechnology for High Performance Construction Materials*, Ed., Schmidt, M., Fehling, E., Glotzbach, C., Fröhlich, S., and Piotrowski, S., Kassel University Press, Kassel, Germany, 2012, pp. 533–540.

[198] Hossain, K., et al., "Bond Strength and Development Length of GFRP Bars in Ultra-High Performance Concrete," Final report, Ontario Ministry of Transportation Research Project, Department of Civil Engineering, Ryerson University, Toronto, Canada, 2011.

[199] Graybeal, B., *Field-Cast UHPC Connections for Modular Bridge Deck Elements, TechBrief*, FHWA-HRT-11-022, Federal Highway Administration, McLean, VA, 2010.

[200] Ruiz, E.D. et al., "Preliminary Investigation Into the Use of UHPC in Prestressed Members," *Proceedings of the PCI National Bridge Conference*, October 22–24, 2007, Phoenix, AZ, Compact Disc, Paper 53.

[201] Ruiz, E.D. et al., "Transfer and Development Lengths of Prestressed Beams Cast With Ultra-High Performance Concrete," *Proceedings of the PCI National Bridge Conference*, October 4–7, 2008, Orlando, FL, Compact Disc, Paper 22.

[202] Bertram, G. and Hegger, J., "Bond Behavior of Strands in UHPC—Tests and Design," *Proceedings of Hipermat 2012 3rd International Symposium on UHPC and Nanotechnology for High Performance Construction Materials*, Ed., Schmidt, M., Fehling, E., Glotzbach, C., Fröhlich, S., and Piotrowski, S., Kassel University Press, Kassel, Germany, 2012, pp. 525–532.

[203] Graybeal, B., "UHPC for Structural Connections," ACI Convention Presentation: UHPCExperience and Developments, October 22, 2012. Available at http://www.concrete.org/Convention/fall-Convention/PresentationDetail.asp?EventId=ZSES S14 [Cited December 14, 2012].

[204] Steinberg, E. and Lubbers, A., "Bond of Prestressing Strands in UHPC," *Proceedings of the 3rd International Symposium on High Performance Concrete/PCI National Bridge Conference*, October 19–22, 2003, Orlando, FL, Compact Disc, Paper 12.

[205] Lubbers, A.R., "Bond Performance Between Ultra-High Performance Concrete and Prestressing Strands," Master's Thesis, College of Engineering and Technology, Ohio University, 2003.

[206] Hegger, J. and Bertram, G., "Anchorage Behavior of Pretensioned Strands in Steel Fiber Reinforced UHPC," *Proceedings of the Second International Symposium on Ultra High Performance Concrete*, Ed., Fehling, E., Schmidt, M., and Stürwald. S., Kassel University Press, Kassel, Germany, 2008, pp. 537–544.

[207] Chen, L. and Graybeal, B., "Finite Element Analysis of Ultra-High Performance Concrete: Modeling Structural Performance of an AASHTO Type II Girder and a 2nd Generation Pi-Girder," FHWA, U.S. Department of Transportation, Report No. FHWA-HRT-11-020, National Technical Information Service Accession No. PB2011-100864, 2010.

[208] Chen, L. and Graybeal, B., "Modeling Structural Performance of Ultra-High Performance Concrete I-Girders," *ASCE Journal of Bridge Engineering*. Vol. 17, No. 5, September–October 2012, pp. 754–764.

[209] Chen, L. and Graybeal, B., "Modeling Structural Performance of Second-Generation Ultrahigh Performance Concrete Pi-Girders," *ASCE Journal of Bridge Engineering*. Vol. 17, No. 4, July–August 2012, pp. 634–643.

[210] Gowripalan, N. and Gilbert, R.I., "Design Guidelines for Ductal Prestressed Concrete Beams," The University of New South Wales, Sydney, Australia, 2000.

[211] Walraven, J., "On the Way to International Design Recommendations for Ultra High Performance Fibre Reinforced Concrete," *Proceedings of Hipermat 2012 3rd International Symposium on UHPC and Nanotechnology for High Performance Construction Materials*, Ed., Schmidt, M., Fehling, E., Glotzbach, C., Fröhlich, S., and Piotrowski, S., Kassel University Press, Kassel, Germany, 2012, pp. 51–58.

[212] *Australian Standard: Concrete Structures, AS3600-94*, Second Edition, Standards Association of Australia, Homebush, NSW, 1996.

[213] *Standard Specifications for Concrete Structures*, Japan Society of Civil Engineers, Tokyo, 2002.

[214] Resplendino, J., "State of the Art of Design and Construction of UHPFRC Structures in France," *Proceedings of Hipermat 2012 3rd International Symposium on UHPC and

Nanotechnology for High Performance Construction Materials, Ed., Schmidt, M., Fehling, E., Glotzbach, C., Fröhlich, S., and Piotrowski, S., Kassel University Press, Kassel, Germany, 2012, pp. 27–41.

[215] Standard Method of Test for Electrical Indication of Concrete's Ability to Resist Chloride Ion Penetration, AASHTO Designation: T 277, *Standard Specifications for Transportation Materials and Methods of Sampling and Testing*, Part 2B—Tests, American Association of State Highway and Transportation Officials, Washington, DC, 2005.

[216] Standard Test Method for Electrical Indication of Concrete's Ability to Resist Chloride Ion Penetration, ASTM C1202, ASTM International, Volume 04.02, West Conshohocken, PA, 2012.

[217] Thomas, M. et al., "Marine Performance of UHPC at Treat Island," *Proceedings of Hipermat 2012 3rd International Symposium on UHPC and Nanotechnology for High Performance Construction Materials*, Ed., Schmidt, M., Fehling, E., Glotzbach, C., Fröhlich, S., and Piotrowski, S., Kassel University Press, Kassel, Germany, 2012, pp. 365–370.

[218] Standard Method of Test for Resistance of Concrete to Chloride Ion Penetration, AASHTO Designation: T 259, *Standard Specifications for Transportation Materials and Methods of Sampling and Testing*, Part 2—Tests, Twenty First Edition, American Association of State Highway and Transportation Officials, Washington, DC, 2002.

[219] Andrade, M.C., Frías, M., and Aarup, B., "Durability of Ultra High Strength Concrete: Compact Reinforced Composite (CRC)," *Proceedings of the Fourth International Symposium on the Utilization of High-Strength/High-Performance Concrete*, 29–31 May 1996, Paris, France, Ed., de Larrard, F. and Lacroix, R., Vol. 2, pp. 529–534.

[220] Piérard, J., Dooms, B., and Cauberg, N., "Evaluation of Durability Parameters of UHPC Using Accelerated Lab Tests," *Proceedings of Hipermat 2012 3rd International Symposium on UHPC and Nanotechnology for High Performance Construction Materials*, Ed., Schmidt, M., Fehling, E., Glotzbach, C., Fröhlich, S., and Piotrowski, S., Kassel University Press, Kassel, Germany, 2012, pp. 371–376.

[221] Graybeal, B.A., "Simultaneous Structural and Environmental Loading of an Ultra-High Performance Concrete Component," FHWA, U.S. Department of Transportation, Report No. FHWA-HRT-10-054. Available through National Technical Information Service at http://www.ntis.gov with Accession No. PB2010110331 [Cited November 23, 2011].

[222] Charron, J.-P., Denarié, E., and Brühwiler, E., "Permeability of Ultra High Performance Fiber Reinforced Concretes (UHPFRC) Under High Stresses," *Materials and Structures*, Vol. 40, No. 3, 2007, pp. 269–277.

[223] Standard Method of Test for Resistance of Concrete to Rapid Freezing and Thawing, AASHTO Designation: T 161, *Standard Specifications for Transportation Materials and Methods of Sampling and Testing*, Part 2A—Tests, American Association of State Highway and Transportation Officials, Washington, DC, 2008.

[224] Müller, C. et al, "Durability of Ultra-High Performance Concrete (UHPC)," *Proceedings of the Third International fib Congress and Exhibition Incorporating the PCI Annual Convention and National Bridge Conference*, Washington, DC, May 29–June 2, 2010, Compact Disc, Paper 135.

[225] Standard Test Method for Scaling Resistance of Concrete Surfaces Exposed to Deicing Chemicals, ASTM C672, ASTM International, Volume 04.02, West Conshohocken, PA, 2012.

[226] Cwirzen, A., Habemehl-Cwirzen, K., and Penttala, V., "The Effect of Heat Treatment on the Salt Freeze-Thaw Durability of UHSC," *Proceedings of the Second International Symposium on Ultra High Performance Concrete*, Ed., Fehling, E., Schmidt, M., and Stürwald. S., Kassel University Press, Kassel, Germany, 2008, pp. 221–230.

[227] Kosmatka, S.H. and Wilson, M.L., *Design and Control of Concrete Mixtures,* Fifteenth Edition, Engineering Bulletin 001, Portland Cement Association, 2011.

[228] Standard Test Method for Abrasion Resistance of Concrete or Mortar Surfaces by the Rotating-Cutter Method, ASTM C944, ASTM International, Volume 04.02, West Conshohocken, PA, 2012.

[229] Hime, W.G., "HPC Tests—Alkali-Silica Reactivity," *HPC Bridge Views*, Issue No. 36 November–December 2004, p. 3.

[230] Standard Test Method for Potential Alkali-Reactivity of Aggregates (Mortar-Bar Method), ASTM C1260, ASTM International, Volume 04.02, West Conshohocken, PA, 2007.

[231] *Fire Resistance Tests—Elements of Building Construction*, ISO 834, International Organization for Standardization, Geneva, Switzerland, 2012.

[232] Heinz, D., Dehn, F., and Urbonas, L., "Fire Resistance of Ultra High Performance Concrete (UHPC)—Testing of Laboratory Samples and Columns under Load," *Proceedings of the International Symposium on Ultra High Performance Concrete*, Ed., Schmidt, M., Fehling, E., and Geisenhanslüke, C., Kassel University Press, Kassel, Germany, 2004, pp. 703–715.

[233] *Fire Behaviour of Building Materials and Building Components; Building Components; Definitions; Requirements and Tests,* DIN 4102-2, Deutsches Institut für Normung E.V., Berlin, Germany, 1977.

[234] Way, R. and Wille, K., "Material Characterization of an Ultra High-Performance-Fibre Reinforced Concrete Under Elevated Temperatures," *Proceedings of Hipermat 2012 3rd International Symposium on UHPC and Nanotechnology for High Performance Construction Materials*, Ed., Schmidt, M., Fehling, E., Glotzbach, C., Fröhlich, S., and Piotrowski, S., Kassel University Press, Kassel, Germany, 2012, pp. 565–572.

[235] Hosser, D., Kampmeier, B., and Hollmann, D., "Behavior of Ultra High Performance Concrete (UHPC) in Case of Fire," *Proceedings of Hipermat 2012 3rd International Symposium on UHPC and Nanotechnology for High Performance Construction Materials*, Ed., Schmidt, M., Fehling, E., Glotzbach, C., Fröhlich, S., and Piotrowski, S., Kassel University Press, Kassel, Germany, 2012, pp. 573–582.

[236] Pimienta, P. et al., "Literature Review on the Behavior of UHPFRC at High Temperature," *Proceedings of Hipermat 2012 3rd International Symposium on UHPC and Nanotechnology for High Performance Construction Materials*, Ed., Schmidt, M., Fehling, E., Glotzbach, C., Fröhlich, S., and Piotrowski, S., Kassel University Press, Kassel, Germany, 2012, pp. 549–556.

[237] Bierwagen, D. and Abu-Hawash, A., "Ultra High Performance Concrete Highway Bridge," *Proceedings of the 2005 Mid-Continent Transportation Research Symposium*, Ames, IA, August 2005.

[238] Endicott, W.A., "A Whole New Cast," *ASPIRE*, Summer 2007, pp. 26–29. Available at http://www.aspirebridge.org [Cited November 23, 2011].

[239] Keierleber, B. et al., "FHWA, Iowa Optimize Pi Girder," *ASPIRE*, Winter 2010, pp. 24–26. Available at http://www.aspirebridge.org [Cited November 23, 2011].

[240] Shutt, C.A., "UHPC Joint Provides New Solutions," *ASPIRE*, Fall 2009, pp. 28–30. Available at http://www.aspirebridge.org. [Cited November 23, 2011].

[241] Royce, M.C., "Concrete Bridges in New York State," *ASPIRE*, Fall 2011, pp. 46–48. Available at http://www.aspirebridge.org [Cited November 23, 2011].

[242] Moore, B., "Little Cedar Creek Bridge—Big Innovation," *ASPIRE*, Spring 2012, p. 27. Available at http://www.aspirebridge.org [Cited April 20, 2012].

[243] Bornstedt, G. and Shike, C., "Connecting Precast Prestressed Concrete Bridge Deck Panels with Ultra High Performance Concrete," *Proceedings of the PCI National Bridge Conference*, October 22–26, 2011, Salt Lake City, UT, Compact Disc, Paper 106.

[244] Anon [Internet], "North American Ductal® Bridge Projects." Available at www.ductal-lafarge.com [Cited January 3, 2013].

[245] Perry, V., Scalzo, P., and Weiss, G., "Innovative Field Cast UHPC Joints for Precast Deck Panel Bridge Superstructures—CN Overhead Bridge at Rainy Lake, Ontario," *Proceedings of the PCI National Bridge Conference*, October 22–24, 2007, Phoenix, AZ, Compact Disc, Paper 3.

[246] Perry, V.H. and Seibert, P.J., "The Use of UHPFRC (Ductal®) for Bridges in North America: The Technology, Applications and Challenges Facing Commercialization," *Proceedings of the Second International Symposium on Ultra High Performance Concrete*, Ed., Fehling, E., Schmidt, M., and Stürwald. S., Kassel University Press, Kassel, Germany, 2008, pp. 815–822. Also available at http://www.ductal-lafarge.com/SLib/17- UHPC%20Paper%20-%20FINAL.pdf [Cited November 23, 2011].

[247] Perry, V.H. and Seibert, P.J., "New Applications of Field Cast UHPC Connections for Precast Bridges," *Proceedings of the PCI National Bridge Conference*, October 22–26, 2011, Salt Lake City, UT, Compact Disc, Paper 101.

[248] Young W. F. and Bopari, J., "Whiteman Creek Bridge—A Synthesis of Accelerated Bridge Construction (ABC), Ultra High Performance Concrete (UHPC), and Fiber Reinforced Polymer Concrete (FRP)," *Proceedings of the PCI National Bridge Conference*, October 22–26, 2011, Salt Lake City, UT, Compact Disc, Paper 111.

[249] Young, W.F. et al., "Whitman Creek Bridge—A Synthesis of Ultra High Performance Concrete and Fiber Reinforced Polymers for Accelerated Bridge Construction," *Proceedings of Hipermat 2012 3rd International Symposium on UHPC and Nanotechnology for High Performance Construction Materials*, Ed., Schmidt, M., Fehling, E., Glotzbach, C., Fröhlich, S., and Piotrowski, S., Kassel University Press, Kassel, Germany, 2012, pp. 849–856.

[250] Keierleber, B. et al., "Design of Buchanan County, Iowa, Bridge, Using Ultra High-Performance Concrete and PI-Beam Cross Section," *Proceedings of the PCI National Bridge Conference*, October 4–7, 2008, Orlando, FL, Compact Disc, Paper 27.

[251] Graybeal, B., *Construction of Field-Cast Ultra-High Performance Concrete Connections, TechNote*, FHWA-HRT-12-038, Federal Highway Administration, McLean, VA, 2012.

[252] Freytag, B. et al., "WILD-Bridge Scientific Preparation for Smooth Realisation," *Proceedings of Hipermat 2012 3rd International Symposium on UHPC and Nanotechnology for High Performance Construction Materials*, Ed., Schmidt, M., Fehling, E., Glotzbach, C., Fröhlich, S., and Piotrowski, S., Kassel University Press, Kassel, Germany, 2012, pp. 881–888.

[253] Hecht, M., "Practical Use of Fibre-Reinforced UHPC in Construction—Production of Precast Elements for Wild-Brücke in Völkermarkt," *Proceedings of Hipermat 2012 3rd International Symposium on UHPC and Nanotechnology for High Performance Construction Materials*, Ed., Schmidt, M., Fehling, E., Glotzbach, C., Fröhlich, S., and Piotrowski, S., Kassel University Press, Kassel, Germany, 2012, pp. 889–896.

[254] Čandrlić, V., Bleiziffer, J., and Mandić, A., "Bakar Bridge Designed in Reactive Powder Concrete," *Proceedings of the Third International Arch Bridges Conference*, 2001, Paris, France, Ed., Abdunur, C., pp. 695–700.

[255] Behloul, M. and Batoz, J.-F. "Ductal® Applications Over the Last Olympiad," *Proceedings of the Second International Symposium on Ultra High Performance Concrete*, Ed., Fehling, E., Schmidt, M., and Stürwald. S., Kassel University Press, Kassel, Germany, 2008, pp. 855–862.

[256] Hajar, Z. et al., "Design and Construction of the World First Ultra-High Performance Concrete Road Bridges," *Proceedings of the International Symposium on Ultra High Performance Concrete*, Ed., Schmidt, M., Fehling, E., and Geisenhanslüke, C., Kassel University Press, Kassel, Germany, 2004, pp. 39–48.

[257] Resplendino, J., "Ultra-High Performance Concretes—Recent Realizations and Research Programs on UHPFRC Bridges in France," *Proceedings of the Second International Symposium on Ultra High Performance Concrete*, Ed., Fehling, E., Schmidt, M., and Stürwald. S., Kassel University Press, Kassel, Germany, 2008, pp. 31–43.

[258] de Matteis, D. et al., "A Fifth French Bridge Including UHPFRC Components, the Widening of the Pinel Bridge, in Rouen, (France)," *Proceedings of the Second International Symposium on Ultra High Performance Concrete*, Ed., Fehling, E., Schmidt, M., and Stürwald. S., Kassel University Press, Kassel, Germany, 2008, pp. 795–802.

[259] Behloul, M. et al., "Ductal® Pont du Diable Footbridge, France," *Tailor Made Concrete Structures*, Ed., Walraven, J. and Stoelhorst, D., 2008, pp. 335–340.

[260] Fehling, E. et al., "The 'Gärtnerplatzbrücke' Design of First Hybrid UHPC-Steel Bridge across the River Fulda in Kassel, Germany," *Proceedings of the Second International Symposium on Ultra High Performance Concrete*, Ed., Fehling, E., Schmidt, M., and Stürwald. S., Kassel University Press, Kassel, Germany, 2008, pp. 581–588.

[261] Fehling, E. et al., "Ultra High Performance Composite Bridge Across the River Fulda in Kassel—Conceptual Design, Design Calculations and Invitation to Tender," *Proceedings of the International Symposium on Ultra High Performance Concrete*, Ed., Schmidt, M., Fehling, E., and Geisenhanslüke, C., Kassel University Press, Kassel, Germany, 2004, pp. 69–75.

[262] Meda, A. and Rosati, G., "Design and Construction of a Bridge in Very High Performance Fiber-Reinforced Concrete," *ASCE Journal of Bridge Engineering*, Vol. 8, No.5, 2003, pp. 281–287.

[263] Buitelaar, P., "Heavy Reinforced Ultra High Performance Concrete," *Proceedings of the International Symposium on Ultra High Performance Concrete*, Ed., Schmidt, M., Fehling, E., and Geisenhanslüke, C., Kassel University Press, Kassel, Germany, 2004, pp. 25–35.

[264] Yuguang, Y., Walraven, J., and den Uijl, J., "Study on Bending Behavior of an UHPC Overlay on a Steel Orthotropic Deck," *Proceedings of the Second International Symposium on Ultra High Performance Concrete*, Ed., Fehling, E., Schmidt, M., and Stürwald. S., Kassel University Press, Kassel, Germany, 2008, pp. 639–646.

[265] Kaptijn, N. and Blom, J., "A New Bridge Deck for the Kaag Bridges—The First CRC (Compact Reinforced Composite) Application in Civil Infrastructure," *Proceedings of the International Symposium on Ultra High Performance Concrete*, Ed., Schmidt, M., Fehling, E., and Geisenhanslüke, C., Kassel University Press, Kassel, Germany, 2004, pp. 49–57.

[266] Šajna, A., Denarié, E., and Bras, V., "Assessment of a UHPFRC Based Bridge Rehabilitation in Slovenia, Two Years After Application," *Proceedings of Hipermat 2012 3rd International Symposium on UHPC and Nanotechnology for High Performance Construction Materials*, Ed., Schmidt, M., Fehling, E., Glotzbach, C., Fröhlich, S., and Piotrowski, S., Kassel University Press, Kassel, Germany, 2012, pp. 937–944.

[267] Resplendino, J., "First Recommendations for Ultra-High-Performance Concretes and Examples of Application," *Proceedings of the International Symposium on Ultra High Performance Concrete*, Ed., Schmidt, M., Fehling, E., and Geisenhanslüke, C., Kassel University Press, Kassel, Germany, 2004, pp. 79–90.

[268] Brühwiler, E. and Denarié, E., "Rehabilitation of Concrete Structures Using Ultra-High Performance Fibre Reinforced Concrete," *Proceedings of the Second International Symposium on Ultra High Performance Concrete*, Ed., Fehling, E., Schmidt, M., and Stürwald. S., Kassel University Press, Kassel, Germany, 2008, pp. 895–902.

[269] Rebentrost, M. and Wight, G., "Experience and Applications of Ultra-High Performance Concrete in Asia," *Proceedings of the Second International Symposium on Ultra High Performance Concrete*, Ed., Fehling, E., Schmidt, M., and Stürwald. S., Kassel University Press, Kassel, Germany, 2008, pp. 19–30.

[270] Anon [Internet], "Shepherd's Creek Ductal Bridge, Australia," Ref: 1755. Available at http://www.vsl.com, System Products Technologies, Ductal®, References [Cited November 21, 2011].

[271] Cavill, B. and Chirgwin, G., "The World's First RPC Road Bridge Shepherds Gully Creek Bridge, NSW," *Proceedings of the 21st Biennial Conference of Concrete Institute of Australia*, Brisbane, Australia, 2003, pp. 89–98.

[272] Anon [Internet], "Ductal Panels for Southern Link Upgrade—Australia," Ref: 2691. Available at http://www.vsl.com, System Products Technologies, Ductal®, References [Cited April 13, 2012].

[273] Okuma, H. et al., "The First Highway Bridge Applying Ultra High Strength Fiber Reinforced Concrete in Japan," *7th International Conference on Short and Medium Span Bridges*, Montreal, Canada, 2006.

[274] Matsubara et al., "Application of a New Type of Ultra High Strength Fiber Reinforced Concrete to a Prestressed Concrete Bridge," *Proceedings of the Second International*

Symposium on Ultra High Performance Concrete, Ed., Fehling, E., Schmidt, M., and Stürwald. S., Kassel University Press, Kassel, Germany, 2008, pp. 787–802.

[275] Resplendino, J. and Petitjean, J., "Ultra-High-Performance Concrete: First Recommendations and Examples of Application," *Proceedings of the 3rd International Symposium on High Performance Concrete/PCI National Bridge Conference*, October 19–22, 2003, Orlando, FL, Compact Disc, Paper 77.

[276] Tanaka, Y. et al., "Innovation and Application of UFC Bridges in Japan," *Proceedings of the International Workshop on Ultra High Performance Fibre Reinforced Concrete—Designing and Building with UHPFRC: State of the Art Development*, 17–19 November 2009, Marseille, France, AFGC/*fib*, Paper 3.1.4.

[277] Musha, H. et al., "Mikaneike Bridge—The First Continuous Girder Bridge Using Ultra-High Strength Fiber Reinforced Concrete," *Journal of Prestressed Concrete*, Japan, Vol. 49, No. 5, September–October, 2007, pp. 18–26.

[278] Lei, V.Y., "Construction of Malaysia's First 50-m-Long UHPdC Composite Road Bridge," *Concrete Technology*, Vol. 4, 2011, pp. 122–127.

[279] Anon [Internet], "Papatoetoe Ductal Footbridge—New Zealand," Ref: 1450. Available at http://www.vsl.com, System Products Technologies, Ductal®, References [Cited November 21, 2011].

[280] Astarlioglu, A., Krauthammer, T., and Felice, C., "State-of-the-Art Report on Fiber Reinforced Ultra-High Performance Concrete," University of Florida, Report CIPPS-TR-003-2010, July 2010.

[281] Parant, E. et al., "Strain Rate Effect on Bending Behavior of New Ultra-High-Performance Cement-Based Composite," *ACI Materials Journal*, Vol. 104, No. 5, September–October 2007, pp. 458–463.

[282] Ngo, T., Mendis, P., and Krauthammer, T., "Behavior of Ultrahigh-Strength Prestressed Concrete Panels Subjected to Blast Loading," *Journal of Structural Engineering*, Vol. 133, No. 11, 2007, pp. 1,582–1,590.

[283] Millard, A. et al, "Dynamic Enhancement of Blast-Resistant Ultra High Performance Fiber-Reinforced Concrete Under Flexural and Shear Loading," *International Journal of Impact Engineering*, Vol. 37, No. 4, 2010, pp. 405–413.

[284] Habel, K. and Gauvreau, P., "Response of Ultra-High Performance Fiber Reinforced Concrete (UHPFRC) to Impact and Static Loading," *Cement and Concrete Composites*, Vol. 30, No. 10, 2008, pp. 938–946.

[285] Millon, O. et al., "Failure Mechanisms of UHPC Components under Explosive Loading," *Proceedings of the Hipermat 2012 3rd International Symposium on UHPC and Nanotechnology for High Performance Construction Materials*, Ed., Schmidt, M., Fehling, E., Glotzbach, C., Fröhlich, S., and Piotrowski, S., Kassel University Press, Kassel, Germany, 2012, pp. 583–591.

[286] Wu, C.et al., "Blast Testing of Ultra High Performance Fibre and FRP-Retrofitted Concrete Slabs," *Engineering Structures*, Vol. 31, No. 9, 2009, pp. 2,060–2,069.

[287] Rebentrost, M., and Wight, G., "Behaviour and Resistance of Ultra High Performance Concrete to Blast Effects," *Proceedings of the Second International Symposium on Ultra-High Performance Concrete*, Ed., Fehling, E., Schmidt, M., and Stürwald. S., Kassel University Press, Kassel, Germany, 2008, pp. 735–742.

[288] Rebentrost, M. and Wight, G., "Investigation of UHPFRC Slabs Under Blast Loads," *Proceedings of the International Workshop on Ultra High Performance Fibre*

Reinforced Concrete—Designing and Building with UHPFRC: State of the Art Development, 17–19 November 2009, Marseille, France, AFGC/*fib*, Paper 4.1.6.

[289] Nöldgen, M. et al, "Ultra High Performance Concrete Structures Under Aircraft Engineer Missile Impact," *Proceedings of the Hipermat 2012 3rd International Symposium on UHPC and Nanotechnology for High Performance Construction Materials*, Ed., Schmidt, M., Fehling, E., Glotzbach, C., Fröhlich, S., and Piotrowski, S., Kassel University Press,, Kassel, Germany, 2012, pp. 593–600.

[290] Almansour, H. and Lounis Z., "Structural Performance of Precast Prestressed Bridge Girders Built With Ultra High Performance Concrete," *Proceedings of the Second International Symposium on Ultra High Performance Concrete*, Ed., Fehling, E., Schmidt, M., and Stürwald. S., Kassel University Press, Kassel, Germany, 2008, pp. 823–830.

[291] *Canadian Highway Bridge Design Code,* CAN/CSA-S6-06, Canadian Standards Association, Mississauga, Ontario, 2006.

[292] Obata, H. et al., "Effect of Improved Interfacial Bond on Whitetopping Using Ultra High Strength Fiber Reinforced Concrete," *Proceedings of the Second International Symposium on Ultra High Performance Concrete*, Ed., Fehling, E., Schmidt, M., and Stürwald. S., Kassel University Press, Kassel, Germany, 2008, pp. 243–250.

[293] Oesterlee, C., Sadouki, H., and Brühwiler, E., "Structural Analysis of a Composite Bridge Girder Combining UHPFRC and Reinforced Concrete," *Proceedings of the Second International Symposium on Ultra High Performance Concrete*, Ed., Fehling, E., Schmidt, M., and Stürwald. S., Kassel University Press, Kassel, Germany, 2008, pp. 647–654.

[294] Schäfers, M. and Seim, W., "Development of Adhesive-Bound UHPC-Timber Composites," *Proceedings of the Second International Symposium on Ultra High Performance Concrete*, Ed., Fehling, E., Schmidt, M., and Stürwald. S., Kassel University Press, Kassel, Germany, 2008, pp. 259–266.

[295] Marchand, P. et al., "Modelling Flexural Tests on UHPFRC Thin-Walled Structures," *Proceedings of Hipermat 2012 3rd International Symposium on UHPC and Nanotechnology for High Performance Construction Materials*, Ed., Schmidt, M., Fehling, E., Glotzbach, C., Fröhlich, S., and Piotrowski, S., Kassel University Press, Kassel, Germany, 2012, pp. 757–764.

[296] Vande Voort, T., Suleiman, M., and Sritharan, S. "Design and Performance Verification of UHPC Piles for Deep Foundations," Final Report, Iowa Highway Research Board Project TR-558, Iowa State University, Ames, IA, 2008.

[297] Ibuk, H. and Beckhaus, K., "Ultra High Performance Concrete for Drill Bits in Special Foundation Engineering," *Proceedings of Hipermat 2012 3rd International Symposium on UHPC and Nanotechnology for High Performance Construction Materials*, Ed., Schmidt, M., Fehling, E., Glotzbach, C., Fröhlich, S., and Piotrowski, S., Kassel University Press, Kassel, Germany, 2012, pp. 807–810.

[298] Schmidt, M., Braun, T., and Möller, H., "Sewer Pipers and UHPC—Development of an UHPC With Earth-Moisture Consistency," *Proceedings of Hipermat 2012 3rd International Symposium on UHPC and Nanotechnology for High Performance Construction Materials*, Ed., Schmidt, M., Fehling, E., Glotzbach, C., Fröhlich, S., and Piotrowski, S., Kassel University Press, Kassel, Germany, 2012, pp. 833–840.

[299] Adam, T. and Ma, J., "Development of an Ultra-High Performance Concrete for Precast Spun Concrete Columns," *Proceedings of Hipermat 2012 3rd International Symposium on UHPC and Nanotechnology for High Performance Construction Materials*, Ed., Schmidt, M., Fehling, E., Glotzbach, C., Fröhlich, S., and Piotrowski, S., Kassel University Press, Kassel, Germany, 2012, pp. 841–848.

[300] Mtiller, C. et al., "Ultra-High Performance Spun Concrete Columns With High-Strength Reinforcement," *Proceedings of Hipermat 2012 3rd International Symposium on UHPC and Nanotechnology for High Performance Construction Materials*, Ed., Schmidt, M., Fehling, E., Glotzbach, C., Fröhlich, S., and Piotrowski, S., Kassel University Press, Kassel, Germany, 2012, pp. 733–740.

[301] Sritharan, S. et al., "Current Research on Ultra High Performance Concrete (UHPC) for Bridge Applications in Iowa," *Proceedings of Hipermat 2012 3rd International Symposium on UHPC and Nanotechnology for High Performance Construction Materials*, Ed., Schmidt, M., Fehling, E., Glotzbach, C., Fröhlich, S., and Piotrowski, S., Kassel University Press, Kassel, Germany, 2012, pp. 857–864.

[302] Shann, S.V. et al., "Application of Ultra-High Performance Concrete (UHPC) as a Thin-Topped Overlay for Concrete Bridge Decks," *Proceedings of Hipermat 2012 3rd International Symposium on UHPC and Nanotechnology for High Performance Construction Materials*, Ed., Schmidt, M., Fehling, E., Glotzbach, C., Fröhlich, S., and Piotrowski, S., Kassel University Press, Kassel, Germany, 2012, pp. 929–936.

[303] Schmidt, C. and Schmidt, M., "'Whitetopping' of Asphalt and Concrete Pavements with Thin Layers of Ultra-High-Performance Concrete—Construction and Economic Efficiency," *Proceedings of Hipermat 2012 3rd International Symposium on UHPC and Nanotechnology for High Performance Construction Materials*, Ed., Schmidt, M., Fehling, E., Glotzbach, C., Fröhlich, S., and Piotrowski, S., Kassel University Press, Kassel, Germany, 2012, pp. 921–927.

[304] Park, S.Y. et al., "Structural Performance of Prestressed UHPC Ribbed Deck for Cable-Stayed Bridge," *Proceedings of Hipermat 2012 3rd International Symposium on UHPC and Nanotechnology for High Performance Construction Materials*, Ed., Schmidt, M., Fehling, E., Glotzbach, C., Fröhlich, S., and Piotrowski, S., Kassel University Press, Kassel, Germany, 2012, pp. 873–880.

[305] Hoffmann, S. and Weiher, H., "Innovative Design of Bridge Bearings by the Use of UHPFRC," *Proceedings of Hipermat 2012 3rd International Symposium on UHPC and Nanotechnology for High Performance Construction Materials*, Ed., Schmidt, M., Fehling, E., Glotzbach, C., Fröhlich, S., and Piotrowski, S., Kassel University Press, Kassel, Germany, 2012, pp. 973–980.

[306] Randl, N. et al., "Study on the Application of UHPC for Precast Tunnel Segments," *Proceedings of Hipermat 20123rd International Symposium on UHPC and Nanotechnology for High Performance Construction Materials*, Ed., Schmidt, M., Fehling, E., Glotzbach, C., Fröhlich, S., and Piotrowski, S., Kassel University Press, Kassel, Germany, 2012, pp. 981–988.

[307] Massicotte, B. and Boucher-Proulx, G., "Seismic Retrofitting of Bridge Piers with UHPFRC Jackets," *Proceedings of the International Workshop on Ultra High Performance Fibre Reinforced Concrete—Designing and Building with UHPFRC: State of the Art Development*, 17–19 November 2009, Marseille, France, AFGC/*fib*, Paper 5.1.1, 9 pp. Also *Designing and Building with UHPFRC: State of the Art and*

Development, Ed., Resplendino, J. and Toutlemonde, F., pp. 531–540, Wiley-ISTE, London, 2010.

[308] *AASHTO LRFD Bridge Construction Specifications*, 3rd Edition, American Association of State Highway and Transportation Officials, Washington, DC, 2010.

[309] Ahlborn, T.M. and Steinberg, E.P., "An Overview of UHPC Efforts Through the Working Group in North America," *Proceedings of Hipermat 2012 3rd International Symposium on UHPC and Nanotechnology for High Performance Construction Materials*, Ed., Schmidt, M., Fehling, E., Glotzbach, C., Fröhlich, S., and Piotrowski, S., Kassel University Press, Kassel, Germany, 2012, pp. 43–50.

BIBLIOGRAPHY

Aaleti, S. et al., "Experimental Evaluation of Structural Behavior of Precast UHPC Waffle Bridge Deck Panels and Connections," *Transportation Research Board 90th Annual Meeting Compendium of Papers*, January 23–27, 2011, Washington, DC, Compact Disc, Session 404, Paper 11-2705.

Aaleti, S.R. et al., "Precast UHPC Waffle Deck Panels and Connections for Accelerated Bridge Construction," *Proceedings of the PCI National Bridge Conference*, October 22–26, 2011, Salt Lake City, UT, Compact Disc, Paper 84.

Aarup, B., "CRC—A Special Fibre Reinforced High Performance Concrete," *Proceedings of the First International RILEM Symposium on Advances in Concrete Through Science and Engineering*, Publication pro048, March 21–24, 2004, Evanston, IL, Ed., Weiss, J. et al.

Aarup, B., "CRC—Precast Applications of Fibre Reinforced Ultra High Performance Concrete," *Proceedings of the International Workshop on Ultra High Performance Fibre Reinforced Concrete—Designing and Building with UHPFRC: State of the Art Development*, 17–19 November 2009, Marseille, France, AFGC/*fib*, Paper 4.1.1.

Aarup, B., "Fibre Reinforced High Performance Concrete for Precast Applications," *Proceedings of the Second International Symposium on Prefabrication*, 17–19 May 2000, Helsinki, Finland, Concrete Association of Finland, pp. 173–178.

Aarup, B., Karlsen, J., and Lindström, G., "Fiber Reinforced High Performance Concrete for In-Situ Cast Joints," *Proceedings of the PCI/FHWA/FIB International Symposium on High Performance Concrete*, September 25–27, 2000, Orlando, FL, Ed., Johal, L.S., pp. 379–387.

Acker, P. and Behloul, M., "Ductal® Technology: A Large Spectrum of Properties, a Wide Range of Applications," *Proceedings of the International Symposium on Ultra High Performance Concrete*, Ed., Schmidt, M., Fehling, E., and GeisenhanslLike, C., Kassel University Press, Kassel, Germany, 2004, pp. 11–23.

Acker, P., "Micromechanical Analysis of Creep and Shrinkage Mechanisms," *Proceedings of ConCreep-6@MIT—Creep, Shrinkage, and Durability Mechanics of Concrete and Other Quasi-Brittle Materials*, 20–22 August 2001, Cambridge, MA, Ed., Ulm, F.-J., Bazant, Z.P., and Wittman, F.H., pp. 15–25.

Acker, P., "Why Does Ultrahigh-Performance Concrete (UHPC) Exhibit Such a Low Shrinkage and Such a Low Creep?" *Autogenous Deformation of Concrete*, Publication

No. SP-220, Jensen, O.M., Bentz, D.P., and Lura P., American Concrete Institute, Farmington Hills, MI, 2004, pp. 141–154.

Adam, T. and Ma, J., "Development of an Ultra-High Performance Concrete for Precast Spun Concrete Columns," *Proceedings of Hipermat 2012 3rd International Symposium on UHPC and Nanotechnology for High Performance Construction Materials*, Ed., Schmidt, M., Fehling, E., Glotzbach, C., Fröhlich, S., and Piotrowski, S., Kassel University Press, Kassel, Germany, 2012, pp. 841–848.

Adeline, R. and Behloul, M., "High Ductile Beams Without Passive Reinforcement," *Proceedings of the Fourth International Symposium on the Utilization of High-Strength/High-Performance Concrete*, 29–31 May 1996, Paris, France, Ed., de Larrard, F. and Lacroix, R., Vol. 3, pp. 1,383-1,390.

Adeline, R. and Cheyrezy, M., "The Sherbrooke Footbridge: The First RPC Structure," *La Technique Française du Béton Précontraint,* FIP 98, Amsterdam, May 1998.

Adeline, R., Behloul, M., and Bernier, G., "Reactive Powder Concrete Pretensioned Beams," *Proceedings of the fip Symposium on Post-Tensioned Concrete Structures, Concrete Society,* Publication CS112/113, 1996, pp. 592–606.

Adeline, R., Lachemi, M., and Blais, P., "Design and Behaviour of the Sherbrooke Footbridge," *Proceedings of the First International Symposium on High-Performance and Reactive Powder Concretes*, Sherbrooke, Quebec, Canada, 16–20 August 1998, Ed., Aïtcin, P.-C., and Delagrave, Y., Vol. 3, pp. 89–97.

Ahlborn, T. and Kollmorgen, G., "Impact of Age and Size on the Mechanical Behavior of Ductal®," Report No. CSD-2004-07, Michigan Technological University, 2004.

Ahlborn, T.M. and Steinberg, E.P., "An Overview of UHPC Efforts Through the Working Group in North America," *Proceedings of Hipermat 2012 3rd International Symposium on UHPC and Nanotechnology for High Performance Construction Materials*, Ed., Schmidt, M., Fehling, E., Glotzbach, C., Fröhlich, S., and Piotrowski, S., Kassel University Press, Kassel, Germany, 2012, pp. 43–50.

Ahlborn, T.M. et al., "Durability and Strength Characterization of Ultra-High Performance Concrete Under Variable Curing Regimes," *Proceedings of the Second International Symposium on Ultra High Performance Concrete*, Ed., Fehling, E., Schmidt, M., and Stürwald, S., Kassel University Press, Kassel, Germany, 2008, pp. 197–204.

Ahlborn, T.M. et al., "Ultra-High Performance Concrete—Study Tour 2002," *Proceedings of the 3rd International Symposium on High Performance Concrete/PCI National Bridge Conference*, October 19–22, 2003, Orlando, FL, Compact Disc, Paper 66.

Ahlborn, T.M., Peuse, E.J., and Misson, D.L., "Ultra-High-Performance-Concrete for Michigan Bridges Material Performance—Phase I," Michigan Department of Transportation, Report No. RC-1525, 2008.

Aïtcin, P.C. and Richard, P., "The Pedestrian/Bikeway Bridge of Sherbrooke," *Proceedings of the Fourth International Symposium on the Utilization of High-Strength/High Performance Concrete*, 29–31 May 1996, Paris, France, Ed., de Larrard, F. and Lacroix, R., Vol. 3, pp. 1,399–1,406.

Aïtcin, P.-C. et al., "The Sherbrooke Reactive Powder Concrete Footbridge," *Structural Engineering International*, Vol. 8, No. 2, 1998, pp. 140–144.

Akhnoukh, A.K., "Development of High Performance Precast/Prestressed Bridge Girders," PhD Dissertation, University of Nebraska-Lincoln, December 2008.

Almansour, H. and Lounis, Z., "Design of Prestressed UHPFRC Girder Bridges According to Canadian Highway Bridge Design Code," *Proceedings of the International Workshop on Ultra High Performance Fibre Reinforced Concrete—Designing and Building with UHPFRC: State of the Art Development*, 17–19 November 2009, Marseille, France, AFGC/*fib*, Paper 5.3.3.

Almansour, H. and Lounis, Z., "Innovative Precast Bridge Superstructure Using Ultra High Performance Concrete Girders," *Proceedings of the PCI National Bridge Conference*, October 22–24, 2007, Phoenix, AZ, Compact Disc, Paper 2.

Almansour, H. and Lounis, Z., "Structural Performance of Precast Prestressed Bridge Girders Built With Ultra High Performance Concrete," *Proceedings of the Second International Symposium on Ultra High Performance Concrete*, Ed., Fehling, E., Schmidt, M., and Stürwald. S., Kassel University Press, Kassel, Germany, 2008, pp. 823–830.

Alonso, C. et al., "Microstructural Changes in High and Ultra High Strength Concrete Exposed to High Temperature Environments," *Quality of Concrete Structures and Recent Advances in Concrete Materials & Testing: An International Conference Honoring V. Mohan Malhotra,* Publication No. SP-229, Ed., Helene, P. et al., American Concrete Institute, Farmington Hills, MI, 2005, pp. 289–302.

An, M.-Z., Zhang, L.-J., and Yi, Q.-X., "Size Effect on Compressive Strength of Reactive Powder Concrete," *Journal of China University of Mining and Technology*, Vol. 18, No. 2, 2008, pp. 279–282.

Andrade, M.C., Frías, M., and Aarup, B., "Durability of Ultra High Strength Concrete: Compact Reinforced Composite (CRC)," *Proceedings of the Fourth International Symposium on the Utilization of High-Strength/High-Performance Concrete*, 29–31 May 1996, Paris, France, Ed., de Larrard, F. and Lacroix, R., Vol. 2, pp. 529–534.

Anon [Internet], "Ductal Footbridges—New Zealand," Ref: 2641. Available at http://www.vsl.com, System Products Technologies, Ductal®, References [Cited November 21, 2011].

Anon [Internet], "Ductal Panels for Southern Link Upgrade—Australia," Ref: 2691. Available at http://www.vsl.com, System Products Technologies, Ductal®, References [Cited April 13, 2012].

Anon [Internet], "North American Ductal® Bridge Projects." Available at www.ductal-lafarge.com [Cited January 3, 2013].

Anon [Internet], "Papatoetoe Ductal Footbridge—New Zealand," Ref: 1450. Available at http://www.vsl.com, System Products Technologies, Ductal®, References [Cited November 21, 2011].

Anon [Internet], "Sanderling Drive Pedestrian Overpass, Calgary, Alberta." Available at www.ductal-lafarge.com [Cited April 13, 2012].

Anon [Internet], "Shepherd's Creek Ductal Bridge, Australia," Ref: 1755. Available at http://www.vsl.com, System Products Technologies, Ductal®, References [Cited November 21, 2011].

Anon, "Feasibility Analysis of Ultra High Performance Concrete for Prestressed Concrete Bridge Application," Weldon, B.D., Simons, B., Daniell, K., New Mexico Department of Transportation Research Bureau Project NM09MSC-01, 2009.

Anon, "Ultra-High Performance Concrete: New Solutions for Today's Highway Infrastructure," *Focus*, September 2011, pp. 3–4.

Arafa, M., Shihada, S., and Karmont, M., "Mechanical Properties of Ultra High Performances Concrete Produced in the Gaza Strip," *Asian Journal of Materials Science,* Vol. 2, No. 1, 2010, pp. 1–12.

Astarlioglu, A., Krauthammer, T., and Felice, C., "State-of-the-Art Report on Fiber Reinforced Ultra-High Performance Concrete," University of Florida, Report CIPPS-TR-003-2010, July 2010.

Ay, L., "Curing Tests on Ultra High Strength Plain and Steel Fibrous Cement Based Composites," *Proceedings of the International Symposium on Ultra High Performance Concrete*, Ed., Schmidt, M., Fehling, E., and GeisenhanslLike, C., Kassel University Press, Kassel, Germany, 2004, pp. 695–701.

Baby, F. et al., "A Proposed Flexural Test Method and Associated Inverse Analysis for UHPFRC," *ACI Materials Journal*, Vol. 109, No. 5, September–October 2012, pp. 545–555.

Baby, F. et al., "Shear Resistance of Ultra High Performance Fibre Reinforced Concrete I-Beams," *Proceedings of the 7th International Conference on Fracture Mechanics of Concrete and Concrete Structures*, Jeju, Korea, May 2010, pp. 1,411–1,417.

Baby, F. et al., "Ultimate Shear Strength of Ultra High Performance Fibre Reinforced Concrete Beams," *Proceedings of Hipermat 2012 3rd International Symposium on UHPC and Nanotechnology for High Performance Construction Materials*, Ed., Schmidt, M., Fehling, E., Glotzbach, C., Fröhlich, S., and Piotrowski, S., Kassel University Press, Kassel, Germany, 2012, pp. 485–492.

Bagheri, A.R., Alibabaie, M., and Zanghane, H., "Effect of Confinement by Steel Tube on Ductility of Reactive Powder Concrete," *Proceedings of the Eighth U.S. National Conference on Earthquake Engineering (100th Anniversary Earthquake Conference)*, April 18–22, 2006, San Francisco, CA, Earthquake Engineering Research Institute, Oakland, CA, Compact Disc.

Banta, T.E., "Horizontal Shear Transfer Between Ultra High Performance Concrete and Lightweight Concrete," Master's Thesis, Virginia Polytechnic Institute and State University, 2005.

Barnett, S.J. et al., "Ultra High Performance Fibre Reinforced Concrete for Explosion Resistant Structures," *Proceedings of Concrete Platform International Conference*, Belfast, Northern Ireland, 2007.

Batoz, J.-F. and Behloul, M., "UHPFRC Development on the Last Two Decades: An Overview," *Proceedings of the International Workshop on Ultra High Performance Fibre Reinforced Concrete—Designing and Building with UHPFRC: State of the Art Development*, 17–19 November 2009, Marseille, France, AFGC/*fib*, Paper 1.2 (Belhoul).

Batoz, J.-F. and Rivallain, M., "Ultrahigh-Performance Concrete Contribution to Sustainable Development," *Proceedings of the International Workshop on Ultra High Performance Fibre Reinforced Concrete—Designing and Building with UHPFRC: State of the Art Development*, 17–19 November 2009, Marseille, France, AFGC/*fib*, Paper 6.2.3.

Behloul, M. and Batoz, J.-F., "Ductal® Applications Over the Last Olympiad," *Proceedings of the Second International Symposium on Ultra High Performance Concrete*, Ed., Fehling, E., Schmidt, M., and Stürwald. S., Kassel University Press, Kassel, Germany, 2008, pp. 855–862.

Behloul, M. and Lee, K.C., "Ductal® Seonyu Footbridge," *Structural Concrete*, Vol. 4, No. 4. 2003, p. 195–201.

Behloul, M. and Lee, K.C., "Innovative Footbridge in Seoul—Seonyu Bridge," *Proceedings of the First fib Congress: Concrete Structures in the 21st Century*, 2002, Osaka, Japan, pp. 183–192.

Behloul, M. and Ricciotti, R., "Footbridge of Peace, Seoul," *Proceedings of the International Symposium on the Role of Concrete Bridges in Sustainable Development,* University of Dundee, Scotland, 3–4 September 2003, Ed., Dhir, R.K., Newlands, M.D., and McCarthy, M.J., pp. 63–72.

Behloul, M. et al., "Ductal® Pont du Diable Footbridge, France," *Tailor Made Concrete Structures*, Ed., Walraven, J. and Stoelhorst, D., 2008, pp. 335–340.

Behloul, M. et al., "Ductal®: Ultra High-Performance Concrete Technology With Ductility," *Proceedings of the 6th International RILEM Symposium on Fibre Reinforced Concretes (BEFIB'2004)*, Publication pro039, Ed., di Prisco, M., Felicetti, R., and Plizzari, G.A., 2004, pp. 1,281–1,290.

Behloul, M. et al., "Fatigue Flexural Behavior of Pre-Cracked Specimens of Special UHPFRC," *Seventh International Symposium on the Utilization of High-Strength/High-Performance Concrete*, Vol. II, Publication No. SP-228, Ed., Russell, H.G., American Concrete Institute, Farmington Hills, MI, 2005, pp.1,253–1,268.

Behloul, M. et al., "Fire Resistance of Ductal® Ultra High Performance Concrete," *Proceedings of the First fib Congress*, Osaka, Japan, 2002, pp. 421–430.

Behloul, M., Bayard, O., and Resplendino, J., "Ductal® Prestressed Girders for a Traffic Bridge in Mayenne, France," *7th International Conference on Short and Medium Span Bridges*, August 23-25, 2006, Montreal, Canada, 10 pp. Available at http://www.ductal-lafarge.com/SLib/34_%20Behloul.SPLC%20Ductal%20bridgeVFnum580.pdf [Cited November 23, 2011].

Behloul, M., Bernier, G., and Cheyrezy, M., "Tensile Behavior of Reactive Powder Concrete (RPC)," *Proceedings of the Fourth International Symposium on the Utilization of High-Strength/High-Performance Concrete*, 29–31 May 1996, Paris, France, Ed. de Larrard, F. and Lacroix, R., Vol. 3, pp. 1,375–1,381.

Behloul, M., Etienne, D., and Maillard, M., "Ductal® Seonyu Footbridge TMD for Better Comfort," *Footbridge*, November 20–22, 2002, Paris.

Behloul, M., Lee, K.C., and Etienne, D., "Seonyu Ductal® Footbridge," *Proceedings of the fib Symposium: Concrete Structures: The Challenge of Creativity*, 26–28 April 2004, Avignon, France, AFGC, pp. 238–239.

Benson, S.D.P. and Karihaloo, B.L., "CARDIFRC®—Development and Mechanical Properties. Part III: Uniaxial Tensile Response and Other Mechanical Properties," *Magazine of Concrete Research*, Vol. 57, No. 8, 2005, pp. 433–443.

Bernier, G., Behloul, M., and Roux, N., "Structural Applications Using Ultra High-Strength Fiber Reinforced Concrete," *Structural Applications of Fiber Reinforced Concrete,* Publication No. SP-182, Ed., Banthis, E., Macdonald, G., and Tatnall, P., American Concrete Institute, Farmington Hills, MI, 1999, pp. 69–86.

Bertram, G. and Hegger, J., "Anchorage Behavior of Strands in Ultra-High Performance Concrete," *Proceedings of the PCI National Bridge Conference*, October 22–26, 2011, Salt Lake City, UT, Compact Disc, Paper 47.

Bertram, G. and Hegger, J., "Bond Behavior of Strands in UHPC—Tests and Design," *Proceedings of Hipermat 2012 3rd International Symposium on UHPC and Nanotechnology for High Performance Construction Materials*, Ed., Schmidt, M., Fehling, E., Glotzbach, C., Fröhlich, S., and Piotrowski, S., Kassel University Press, Kassel, Germany, 2012, pp. 525–532.

Bertram, G. and Hegger, J., "Design and Construction with Ultra-High Performance Concrete," *Proceedings of the Third International fib Congress and Exhibition Incorporating the PCI Annual Convention and National Bridge Conference*, Washington, DC, May 29–June 2, 2010, Compact Disc, Paper 704.

Bertram, G. and Hegger, J., "Pretensioned UHPC Beams With and Without Openings," *Proceedings of the Third International fib Congress and Exhibition Incorporating the PCI Annual Convention and National Bridge Conference*, Washington, DC, May 29–June 2, 2010, Compact Disc, Paper 236.

Bertram, G. and Hegger, J., "Shear Behavior of Pretensioned UHPC Beams—Tests and Design," *Proceedings of Hipermat 2012 3rd International Symposium on UHPC and Nanotechnology for High Performance Construction Materials*, Ed., Schmidt, M., Fehling, E., Glotzbach, C., Fröhlich, S., and Piotrowski, S., Kassel University Press, Kassel, Germany, 2012, pp. 493–500.

Bertram, G. and Hegger, J., "Size Effects of Pretensioned Ultra-High Performance Concrete Beams," *Proceedings of the PCI National Bridge Conference*, October 22–26, 2011, Salt Lake City, UT, Compact Disc, Paper 48.

Bétons Fibrés à Ultra-Hautes Performances, Recommandations Provisoires, (*Ultra High Performance Fibre-Reinforced Concretes*, Interim Recommendations), SETRA-AFGC, Groupe de travail BFUP, Paris, France, January 2002. (in French and English). Available at http://www.afgc.asso.fr/statuts-de-lassociation-francaise-de-genie-civil.html [Cited November 23, 2011].

Bierwagen, D. and Abu-Hawash, A., "Ultra High Performance Concrete Highway Bridge," *Proceedings of the PCI National Bridge Conference*, October 17–20, 2004, Atlanta, GA, Compact Disc, Paper 59.

Bierwagen, D. and Abu-Hawash, A., "Ultra High Performance Concrete Highway Bridge," *Proceedings of the 2005 Mid-Continent Transportation Research Symposium,* Ames, IA, August 2005.

Bierwagen, D. and McDonald, N., "Ultra High Performance Concrete Highway Bridge," *Proceedings of the PCI 51st Annual Convention/Exhibition and the PCI National Bridge Conference,* Palm Springs, CA, October 16–19, 2005, Compact Disc.

Bierwagen, D. et al., "Design of Buchanan County, Iowa, Bridge, Using Ultra-High Performance Concrete and Pi-Girder Cross Section," *Proceedings of the Third International fib Congress and Exhibition Incorporating the PCI Annual Convention and National Bridge Conference*, Washington, DC, May 29–June 2, 2010, Compact Disc, Paper 172.

Bierwagen, D. et al., "Ultra-High Performance Concrete in Iowa," *HPC Bridge Views*, Issue No. 57, September/October 2009. Available at http://www.hpcbridgeviews.org [Cited November 23, 2011].

Bierwagen, D. et al., "Ultra-High Performance Concrete Waffle Slab Bridge Deck for Wapello County, Iowa," *HPC Bridge Views*, Issue No. 65, January/February 2011. Available at http://www.hpcbridgeviews.org [Cited November 23, 2011].

Bindiganvile, V., Bantha, N., and Aarup, B., "Impact Response of Ultra-High-Strength Fiber-Reinforced Cement Composite," *ACI Materials Journal*, Vol. 99, No. 6, November–December 2002, pp. 543–548.

Birelli, G., "Seance Introductive—Historique des BFUP et Realizations Marquantes," (UHPFRG: Historical Perspective and Outstanding Realizations), *Proceedings of the International Workshop on Ultra High Performance Fibre Reinforced Concrete—Designing and Building with UHPFRC: State of the Art and Development*, 17–18 November 2009, Marseille, France, AFGC/*fib*, Paper 1.3 (Birelli).

Blaise, P.Y. and Couture, M., "Precast, Prestressed Pedestrian Bridge—World's First Reactive Powder Concrete Structure," *PCI Journal*, Vol. 44, No. 5, September/October 1999, pp. 60–71.

Bonneau, O. et al., "Mechanical Properties and Durability of Two Industrial Reactive Powder Concretes," *ACI Materials Journal*, Vol. 94, No. 4, July–August 1997, pp. 286–290.

Bonneau, O. et al., "Reactive Powder Concretes: From Theory to Practice," *Concrete International*, Vol. 18, No. 4, April 1996, pp. 47–49.

Bornemann, R. and Faber, S., "UHPC With Steel- and Non-Corroding High-Strength Polymer Fibres Under Static and Cyclic Loading," *Proceedings of the International Symposium on Ultra High Performance Concrete*, Ed., Schmidt, M., Fehling, E., and GeisenhanslLike, C., Kassel University Press, Kassel, Germany, 2004, pp. 673–682.

Bornstedt, G. "Connecting Precast Concrete Bridge Deck Panels With Ultra High Performance Concrete (UHPC)," *Western Bridge Engineer's Seminar*, Phoenix, AZ, September 25–28, 2011. Abstract only.

Bornstedt, G. and Shike, C., "Connecting Precast Prestressed Concrete Bridge Deck Panels With Ultra High Performance Concrete," *Proceedings of the PCI National Bridge Conference*, October 22–26, 2011, Salt Lake City, UT, Compact Disc, Paper 106.

Boulay, C., Rossi, P., and Tailhan, J.-L., "Uniaxial Tensile Test on a New Cement Composite Having a Hardening Behaviour," *Sixth RILEM Symposium on Fibre-Reinforced Concretes (FRC), BEFIB 2004*, Publication Pro039, Varenna, Italy, September 20–22, 2004, pp. 61–68.

Brauns, J. and Rocens, K., "Stress State Optimization in Steel-Concrete Composite Elements," *Proceedings of the International Symposium on Ultra High Performance Concrete*, Ed., Schmidt, M., Fehling, E., and GeisenhanslLike, C., Kassel University Press, Kassel, Germany, 2004, pp. 413–423.

Brouwer, G., "Bridge to the Future," *Civil Engineering—ASCE*, Vol. 71, No. 11, November 2001, pp. 50–55.

BrLihwiler, E. and Denarié, E., "Rehabilitation of Concrete Structures Using Ultra-High Performance Fibre Reinforced Concrete," *Proceedings of the Second International Symposium on Ultra High Performance Concrete*, Ed., Fehling, E., Schmidt, M., and StLirwald. S., Kassel University Press, Kassel, Germany, 2008, pp. 895–902.

BrLihwiler, E. et al., "Design of an Innovative Composite Road Bridge Combining Reinforced Concrete with Ultra-High Performance Fiber Reinforced Concrete," *Proceedings of IABSE Symposium Improving Infrastructure Worldwide—Bringing People Closer*, IABSE, Weimar, Germany, Vol. 93, 2007, pp. 80–81.

BrLihwiler, E., Denarié, E., and Habel, K., "Ultra-High Performance Fibre Reinforced Concrete for Advanced Rehabilitation of Bridges," *Proceedings of the fib Symposium—*

Keep Concrete Attractive, 23–25 May 2005, Budapest, Hungary, Ed., Balazs, G.L. and Borosnyoi, A., Vol. 2, pp. 951–956. Available at http://www.dist.unina.it/proc/2005/FIB_SYMPOSIUM/vol_2/951.pdf [Cited November 23, 2011].

Budelmann, H. and Ewert, J., "Mechanical Properties of Ultra-High Performance Concrete (UHPC) at Early Age," *Proceedings of Hipermat 2012 3rd International Symposium on UHPC and Nanotechnology for High Performance Construction Materials*, Ed., Schmidt, M., Fehling, E., Glotzbach, C., Fröhlich, S., and Piotrowski, S., Kassel University Press, Kassel, Germany, 2012, pp. 301–308.

Buitelaar, P. et al., "Reinforced High Performance Concrete for Rehabilitation of Orthotropic Steel Bridge Decks," *Proceedings of the Eleventh International Conference and Exhibition on Structural Faults and Repair*, 13–15 June 2006, Edinburgh, Scotland.

Buitelaar, P., "Heavy Reinforced Ultra High Performance Concrete," *Proceedings of the International Symposium on Ultra High Performance Concrete*, Ed., Schmidt, M., Fehling, E., and GeisenhanslLike, C., Kassel University Press, Kassel, Germany, 2004, pp. 25–35.

Buitelaar, P., "Ultra High Performance Concrete: Developments and Applications During 25 Years," Available at www.ferroplan.com [Cited April 14, 2012].

Bunje, K. and Fehling, E., "About Shear Force and Punching Shear Resistance of Structural Elements of Ultra High Performance Concrete," *Proceedings of the International Symposium on Ultra High Performance Concrete*, Ed., Schmidt, M., Fehling, E., and GeisenhanslLike, C., Kassel University Press, Kassel, Germany, 2004, pp. 401–411.

Burkart, I. and MLiller, H.S., "Creep and Shrinkage Characteristics of Ultra High Strength Concrete (UHPC)," *Proceedings of the Second International Symposium on Ultra High Performance Concrete*, Ed., Fehling, E., Schmidt, M., and StLirwald. S., Kassel University Press, Kassel, Germany, 2008, pp. 469–476.

Cadoni, E., Caverzan, A., and di Prisco, M., "Dynamic Behaviour of HPFR Cementitious Composites," *Proceedings of the Second International Symposium on Ultra High Performance Concrete*, Ed., Fehling, E., Schmidt, M., and Stürwald. S., Kassel University Press, Kassel, Germany, 2008, pp. 743–750.

Camacho, E., López, J.A., and Serna Ros, P., "Definition of Three Levels of Performance for the UHPFRC-VHPFRC with Available Materials," *Proceedings of Hipermat 2012 3rd International Symposium on UHPC and Nanotechnology for High Performance Construction Materials*, Ed., Schmidt, M., Fehling, E., Glotzbach, C., Fröhlich, S., and Piotrowski, S., Kassel University Press, Kassel, Germany, 2012, pp. 249–256.

Campbell Sr., R.L. et al., "Reactive Powder Concrete for Producing Sewer, Culvert, and Pressure Pipes," Technical Report CPAR-SL-98-3, U.S. Army Corps of Engineers, Washington, DC, 1998.

Čandrlić, V., Bjegovic, D., and Skazlic, M., "Croatian Experience With RPC," *Proceedings of the Sixth International Conference on Short and Medium Span Bridges—Developments in Short and Medium Span Bridge Engineering*, 31 July 2–August 2002, Vancouver, British Columbia, Vol. 1, pp. 355–362.

Čandrlić, V., Bleiziffer, J., and Mandić, A., "Bakar Bridge Designed in Reactive Powder Concrete," *Proceedings of the Third International Arch Bridges Conference*, 2001, Paris, France, Ed., Abdunur C., pp. 695–700.

Čandrlić, V., Mandić, A., and Bleiziffer, J., "The Largest Concrete Arch Bridge Designed of RPC 200," *Proceedings of the Fourth Symposium on Strait Crossings* Bergen, Norway, 2–5 September 2001, Ed., Krokeborg, J., pp. 145–151.

Carbonell, M.A. et al., "Bond Strength Between UHPC and Normal Strength Concrete (NSC) in Accordance With Split Prism and Freeze-Thaw Cycling Tests," *Proceedings of Hipermat 2012 3rd International Symposium on UHPC and Nanotechnology for High Performance Construction Materials*, Ed., Schmidt, M., Fehling, E., Glotzbach, C., Fröhlich, S., and Piotrowski, S., Kassel University Press, Kassel, Germany, 2012, pp. 377–384.

Carcasses, M., "Béton Fibrés Ultra Performant (BFUP), Les Nouvelles Recommandations AFGC Chapitre 3 Durabilité des BFUP" (Updated AFGC Recommendations: Section 3 Durability), *Proceedings of the International Workshop on Ultra High Performance Fibre Reinforced Concrete—Designing and Building with UHPFRC: State of the Art Development*, 17–19 November 2009, Marseille, France, AFGC/*fib*, Paper 6.1.4.

Casanova, P. and Durukal, A., "A Fire-Resistant Ultra High Performance Fibre Reinforced Concrete Footbridge without Reinforcement," *Proceedings of the First International Conference on Footbridges*, 20–22 November 2002, Paris.

Cauberg. N. et al., "Shear Capacity of UHPC—Beam Tests," *Proceedings of Hipermat 2012 3rd International Symposium on UHPC and Nanotechnology for High Performance Construction Materials*, Ed., Schmidt, M., Fehling, E., Glotzbach, C., Fröhlich, S., and Piotrowski, S., Kassel University Press, Kassel, Germany, 2012, pp. 451–458.

Cavill, B. and Chirgwin, G., "The World's First RPC Road Bridge Shepherds Gully Creek Bridge, NSW," *Proceedings of the 21st Biennial Conference of Concrete Institute of Australia*, Brisbane, Australia, 2003, pp. 89–98.

Chan, Y.-W. and Chu, S.-H., "Effect of Silica Fume on Steel Fiber Bond Characteristics in Reactive Powder Concrete," *Cement and Concrete Research*, Vol. 34, No. 7, 2004, pp. 1,167–1,172.

Chanvillard, G. and Rigaud, S., "Complete Characterization of Tensile Properties of Ductal® UHPFRC According to the French Recommendations," *Proceedings of the 4th International RILEM Workshop on High Performance Fiber-Reinforced Cement Composites (HPFRCC4)* Publication pro030, Ed., Naaman, A.E. and Reinhardt, H.W., 2003, pp. 2–34.

Charron, J.-P., Denarié, E., and Brühwiler, E., "Permeability of UHPFRC Under High Stresses," *Proceedings of the International RILEM Symposium on Concrete Science and Engineering*, Publication pro048, March 21–24, 2004, Evanston, IL, Ed., Weiss, J. et al.

Charron, J.-P., Denarié, E., and Brühwiler, E., "Permeability of Ultra High Performance Fiber Reinforced Concretes (UHPFRC) Under High Stresses," *Materials and Structures*, Vol. 40, No. 3, 2007, pp. 269–277.

Charron, J.-P., Niamba, E., and Massicotte, B., "Design of Bridge Parapets in High and Ultra High Performance Fibre Reinforced Concretes," *Proceedings of the 7th RILEM International Symposium on Fibre Reinforced Concrete: Design and Applications—BEFIB 2008*, Publication pro060, Ed., Gettu, R., 2008, pp. 951–960.

Charron, J.-P., Niamba, E., and Massicotte, B., "Precast Bridge Parapets in Ultra-High Performance Fibre Reinforced Concrete," *Proceedings of the International Workshop on Ultra High Performance Fibre Reinforced Concrete—Designing and Building with*

UHPFRC: State of the Art Development, 17–19 November 2009, Marseille, France, AFGC/*fib*, Paper 4.2.2.

Chauvel, D. et al., "First Design Rules for RPC Beams," *Proceedings of the First International Symposium on High-Performance and Reactive Powder Concrete*, 16–20 August 1998, Sherbrooke, QC, Ed., Aïtcin, P.-C., and Delagrave, Y., Vol. 3, pp. 1–15.

Chen, B., Du, R., and Savor, Z., "Trial Design and Model Tests on Reactive Powder Concrete (RPC) Arches," *3rd Chinese-Croatian Joint Colloquium Sustainable Arch Bridges*, Zagreb, Croatia, July 2011, pp. 31–44.

Chen, J. and Gilles, C., "UHPC Composites Based on Glass Fibers With High Fluidity, Ductility, and Durability," *Proceedings of Hipermat 2012 3rd International Symposium on UHPC and Nanotechnology for High Performance Construction Materials*, Ed., Schmidt, M., Fehling, E., Glotzbach, C., Fröhlich, S., and Piotrowski, S., Kassel University Press, Kassel, Germany, 2012, pp. 265–272.

Chen, L. and Graybeal, B.A. "Finite Element Analysis of Ultra-High Performance Concrete: Modeling Structural Performance of an AASHTO Type II Girder and a 2nd Generation Pi-Girder," FHWA, U.S. Department of Transportation, Report No. FHWA-HRT-11-020, National Technical Information Service Accession No. PB2011-100864, 2010.

Chen, L. and Graybeal, B.A., "Modeling Structural Performance of Ultra-High Performance Concrete I-Girders," *ASCE Journal of Bridge Engineering*, Vol. 17, No. 5, September–October 2012, pp. 754–764.

Chen, L. and Graybeal, B., "Modeling Structural Performance of Second-Generation Ultrahigh Performance Concrete Pi-Girders," *ASCE Journal of Bridge Engineering*, Vol. 17, No. 4, July–August 2012, pp. 634–643.

Cheyrezy, M. and Behloul, M., "Creep and Shrinkage of Ultra-High Performance Concrete," *Proceedings of the Sixth International Conference, Concreep 6-@MIT: Creep, Shrinkage and Durability Mechanics of Concrete and Other Quasi-Brittle Materials*, 13–15 August 2001, Cambridge, MA, Eds. Ulm, F.-J., et al., (Boston: Pergamon Press, 2001), pp. 527–538.

Cheyrezy, M. et al., "Bond Strength of Reactive Powder Concrete," *Proceedings of the Thirteenth FIP Congress (FIP98)*, 23–29 May 1998, Amsterdam, Vol. 1, pp. 65–68.

Cheyrezy, M., "Structural Applications of RPC," *Proceedings of the International Conference on New Technologies in Structural Engineering*, 2–5 July 1997, Lisbon, Portugal, Vol. 1, pp. 5–14.

Cheyrezy, M., Maret, V., and Frouin, L., "Microstructural Analysis of RPC (Reactive Powder Concrete)," *Cement and Concrete Research*, Vol. 25, No. 7, 1995, pp. 1,491–1,500.

Cizmar, D., Mestrovic, D., and Radić, J., "Arch Bridge Made of Reactive Powder Concrete," *Proceedings of the Third International Conference on High Performance Structures and Materials*, 3–5 May 2006, Ostend, Belgium, Ed., Brebbia, C.A., pp. 429–437.

Cizmar, D., Nizic, A., and Mestrovic, D., "The Reactive Powder Concrete Bridge Across Bakar Strait," *Proceedings of the IABSE Symposium, Responding to Tomorrow's Challenges in Structural Engineering*, 13–15 September 2006, IABSE Vol. 92, pp. 224–225.

Collepardi, S. et al., "Mechanical Properties of Modified Reactive Powder Concrete," *Proceedings of the Fifth CANMET/ACI International Conference on Superplasticizers and Other Chemical Admixtures in Concrete*, 1997, Rome, Italy, Publication No. SP-173, Ed., Malhotra V.M., American Concrete Institute, Farmington Hills, MI, pp. 1–21.

Concrete Institute of Australia Resource Center, "Prestressed Structural Concrete: New Developments and Applications," 2011.

Coppola, L. et al., "Influence of Super-Plasticizer Type on Compressive Strength of Reactive Powder Mortars," *Proceedings of the Fifth CANMET/ACI International Conference on Superplasticizers and Other Chemical Admixtures in Concrete*, 1997, Rome, Italy, Publication No. SP-173, Ed., Malhotra V.M., American Concrete Institute, Farmington Hills, MI, pp. 537–557.

Cousins, T., Roberts-Wollmann, C., and Sotelino, E., "UHPC Deck Panels for Rapid Bridge Construction and Long Term Durability," *Proceedings of the Second International Symposium on Ultra High Performance Concrete*, Ed., Fehling, E., Schmidt, M., and Stürwald. S., Kassel University Press, Kassel, Germany, 2008, pp. 699–705.

Crane, C.K. and Kahn, L.F., "Interface Shear Capacity of Small UHPC/HPC Composite T-Beams," *Proceedings of Hipermat 2012 3rd International Symposium on UHPC and Nanotechnology for High Performance Construction Materials*, Ed., Schmidt, M., Fehling, E., Glotzbach, C., Fröhlich, S., and Piotrowski, S., Kassel University Press, Kassel, Germany, 2012, pp. 459–467.

Crane, C.K., "Shear and Shear Friction of Ultra-High Performance Concrete Bridge Girders," PhD Dissertation, Georgia Institute of Technology, Atlanta, GA, 2010.

Curbach, M. and Speck, K., "Ultra High Performance Concrete Under Biaxial Compression," *Proceedings of the Second International Symposium on Ultra High Performance Concrete*, Ed., Fehling, E., Schmidt, M., and Stürwald. S., Kassel University Press, Kassel, Germany, 2008, pp. 477–484.

Cwirzen, A., Habemehl-Cwirzen, K., and Penttala, V., "The Effect of Heat Treatment on the Salt Freeze-Thaw Durability of UHSC," *Proceedings of the Second International Symposium on Ultra High Performance Concrete*, Ed., Fehling, E., Schmidt, M., and Stürwald. S., Kassel University Press, Kassel, Germany, 2008, pp. 221–230.

D'Alessandro, K.C. et al., "Investigation of Biaxial Stress States of Ultra-High Performance Concrete Bridge Girders Through Small Panel Testing and Finite Element Analysis," *Proceedings of the International Workshop on Ultra High Performance Fibre Reinforced Concrete—Designing and Building with UHPFRC: State of the Art and Development*, 17–18 November 2009, Marseille, France, AFGC/*fib*, Paper P 2.

DAfStb, *State-of-the-Art Report on Ultra-High Performance Concrete—Concrete Technology and Design*, Deutscher Ausschuss fur Stahlbeton, Berlin, Germany, 2005.

Dallaire, E., Aïtcin, P., and Lachemi, M., "High-Performance Powder," *Civil Engineering*, January 1998, pp. 48–51.

Dattatreya, J.K., Harish, K.V., and Neelamegam, M., "Investigation of the Flexural Toughness and Fracture Energy of High and Ultra High Performance Fiber Reinforced Concretes," *Proceedings of the 7th RILEM International Symposium on Fibre Reinforced Concrete: Design and Applications— BEFIB 2008*, Publication pro060, Ed., Gettu, R., 2008, pp. 231–242.

de Larrard, F. and Sedran, T., "Optimization of Ultra-High-Performance Concrete by the Use of a Packing Model," *Cement and Concrete Research*, Vol. 24, No. 6, 1994, pp. 997–1,009.

de Larrard, F., "The High-Performance Cementitious Material (HPCM), a Cousin of UHPFRC for Long-Life Pavement," *Proceedings of the International Workshop on Ultra High Performance Fibre Reinforced Concrete—Designing and Building with UHPFRC: State of the Art Development*, 17–19 November 2009, Marseille, France, AFGC/*fib*, Paper 4.2.5.

de Matteis, D. et al., "A Fifth French Bridge Including UHPFRC Components, the Widening of the Pinel Bridge, in Rouen, (France)," *Proceedings of the Second International Symposium on Ultra High Performance Concrete*, Ed., Fehling, E., Schmidt, M., and Stürwald. S., Kassel University Press, Kassel, Germany, 2008, pp. 795–802.

Degen, B., "Shear Design and Behavior of Ultra-High Performance Concrete," Master's Thesis, Iowa State University, Ames, IA, 2006.

Dehn, F., "Fiber-Reinforced Ultra-High-Performance Concrete in Practice—Precast Bridge in Volkermarkt—Quality Assurance and Support by an Expert Consultant," *Betonwerk und FertigteilTechnik/Concrete Plant and Precast Technology*, Vol. 75 No. 2, 2009.

Delauzun, O. et al., "Construction du Pont de la Chabotte en BFUP sur l'Autoroute A51, (Overpass No. 34 "Viaduct de la Chabotte" Over Motorway A51,)" *Proceedings of the International Workshop on Ultra High Performance Fibre Reinforced Concrete—Designing and Building with UHPFRC: State of the Art Development*, 17–19 November 2009, Marseille, France, AFGC/*fib*, Paper 3.1.3.

Denarié, E. and Brühwiler, E., "Structural Rehabilitations With Ultra-High Performance Fibre Reinforced Concretes (UHPFRC)," *International Journal for Restoration of Buildings and Monuments*, Vol. 12, No. 5/6, 2006, pp. 453–468.

Diederichs, U. and Mertzsch, O., "Behaviour of Ultra High Strength Concrete at High Temperatures," *Proceedings of the Second International Symposium on Ultra High Performance Concrete*, Ed., Fehling, E., Schmidt, M., and Stürwald. S., Kassel University Press, Kassel, Germany, 2008, pp. 347–354.

Dowd, W.M. and Dauriac, C.E., "Development of Reactive Powder Concrete (RPC) Precast Products for the United States Market," *Proceedings of the First International Symposium on High-Performance and Reactive Powder Concretes*, 16–20 August 1998, Sherbrooke, Quebec, Canada, Ed., Aïtcin, P.-C., and Delagrave, Y., Vol. 3, pp. 37–57.

Dowd, W.M. and O'Neil, E.F., "Development of Reactive Powder Concrete (RPC) Precast Products for the USA Market," *Proceedings of the Fourth International Symposium on the Utilization of High-Strength/High-Performance Concrete*, 29–31 May 1996, Paris, France, Ed., de Larrard, F. and Lacroix, R., Vol. 3, pp. 1,391–1,398.

Dowd, W.M., Christophe, E.D., and Adeline, R., "Reactive Powder Concrete for Bridge Construction," *Fifth Materials Engineering Congress Materials and Construction: Exploring the Connection, May 10–12,* American Society of Civil Engineering, New York, 1999, pp. 359–366.

Dowd, W., "Reactive Powder Concrete—Ultra-High Performance Cement Based Composite," 1999 NOVA Award Nomination No. 27, Construction Innovation Forum, 1999. Available at http://www.cif.org [Cited November 23, 2011].

Dugat, J., Roux, N., and Bernier, G., "Mechanical Properties of Reactive Powder Concrete," *Materials and Structures*, Vol. 29, No. 4, 1996, pp. 233–240.

Durukal, A. and Casanova, P., "The Sernaises Footbridge: A Fire-Resistant Ultra High Performance Fibre Reinforced Concrete Footbridge Without Reinforcement," *Footbridge, 1st International Conference*, Paris, November 20–22, 2002.

Elmahdy, A. et al., "Structural Evaluation of Hybrid FRP-UHPC Bridge Girders," *Proceedings of the International Workshop on Ultra High Performance Fibre Reinforced Concrete—Designing and Building with UHPFRC: State of the Art Development*, 17–19 November 2009, Marseille, France, AFGC/*fib*, Paper 4.2.7.

Empelmann, M. and Müller, C., "Concrete Columns—New Possibilities from NPC up to UHPC," *Proceedings of the Third International fib Congress and Exhibition Incorporating the PCI Annual Convention and National Bridge Conference*, Washington, DC, May 29–June 2, 2010, Compact Disc, Paper 413.

Empelmann, M. and Oettel, V., "UHPFRC Box Girders Under Torsion," *Proceedings of Hipermat 2012 3rd International Symposium on UHPC and Nanotechnology for High Performance Construction Materials*, Ed., Schmidt, M., Fehling, E., Glotzbach, C., Fröhlich, S., and Piotrowski, S., Kassel University Press, Kassel, Germany, 2012, pp. 517–524.

Empelmann, M., Teutsch, M., and Steven, G., "Load-Bearing Behaviour of Centrically Loaded UHPFRC Columns," *Proceedings of the Second International Symposium on Ultra High Performance Concrete*, Ed., Fehling, E., Schmidt, M., and Stürwald. S., Kassel University Press, Kassel, Germany, 2008, pp. 521–528.

Endicott, W.A., "A Whole New Cast," *ASPIRE*, Summer 2007, pp. 26–29. Available at http://www.aspirebridge.org [Cited November 23, 2011].

Eppers, S. and Müller, C., "Autogenous Shrinkage Strain of Ultra-High-Performance Concrete (UHPC)," *Proceedings of the Second International Symposium on Ultra High Performance Concrete*, Ed., Fehling, E., Schmidt, M., and Stürwald. S., Kassel University Press, Kassel, Germany, 2008, pp. 433–441.

Fehling, E. and Ismail, M., "Experimental Investigations on UHPC Structural Elements Subject to Pure Torsion," *Proceedings of Hipermat 2012 3rd International Symposium on UHPC and Nanotechnology for High Performance Construction Materials*, Ed., Schmidt, M., Fehling, E., Glotzbach, C., Fröhlich, S., and Piotrowski, S., Kassel University Press, Kassel, Germany, 2012, pp. 501–508.

Fehling, E. and Leutbecher, T., "Tensile Behavior of Ultra-High Performance Concrete (UHPC) Reinforced with a Combination of Steel-Fibers and Rebars," Ed., Zingoni, A., *Proceedings of the 3rd International Conference on Structural Engineering, Mechanics and Computation (SEMC)*, Cape Town, South Africa, 2007, pp. 549–550.

Fehling, E. and Leutbecher, T., "Ultra high Performance Concrete (UHPC)—A Challenge in Structural Design," *Proceedings of the fib Symposium—Keep Concrete Attractive*,

23–25 May 2005, Budapest, Hungary, Ed., Balazs, G.L. and Borosnyoi, A., Vol. 2, pp. 251–256. Available on Compact Disc from http://www.fib-international.org/proceedings-of-previousconferences [Cited November 23, 2011].

Fehling, E. and Thiemicke, J., "Experimental Investigation on I-Shaped UHPC Beams with Combined Reinforcement under Shear Load," *Proceedings of Hipermat 2012 3rd International Symposium on UHPC and Nanotechnology for High Performance Construction Materials*, Ed., Schmidt, M., Fehling, E., Glotzbach, C., Fröhlich, S., and Piotrowski, S., Kassel University Press, Kassel, Germany, 2012, pp. 477–484.

Fehling, E. et al., "The 'Gärtnerplatzbrücke' Design of First Hybrid UHPC-Steel Bridge Across the River Fulda in Kassel, Germany," *Proceedings of the Second International Symposium on Ultra High Performance Concrete*, Ed., Fehling, E., Schmidt, M., and Stürwald. S., Kassel University Press, Kassel, Germany, 2008, pp. 581–588.

Fehling, E. et al., "Ultra High Performance Composite Bridge Across the River Fulda in Kassel— Accompanying Investigations According to the Required Agreement by the Authorities," *Proceedings of the International Symposium on Ultra High Performance Concrete*, Ed.,

Schmidt, M., Fehling, E., and Geisenhanslüke, C., Kassel University Press, Kassel, Germany, 2004, pp. 819–825.

Fehling, E. et al., "Ultra High Performance Composite Bridge Across the River Fulda in Kassel— Conceptual Design, Design Calculations and Invitation to Tender," *Proceedings of the International Symposium on Ultra High Performance Concrete*, Ed., Schmidt, M., Fehling, E., and Geisenhanslüke, C., Kassel University Press, Kassel, Germany, 2004, pp. 69–75.

Fehling, E., Bunje, K, and Schmidt, M., "Gärtnerplatz-Bridge Over River Fulda in Kassel—Multispan Hybrid UHPC—Steel Bridge," *Proceedings of the International Workshop on Ultra High Performance Fibre Reinforced Concrete—Designing and Building with UHPFRC: State of the Art Development*, 17–19 November 2009, Marseille, France, AFGC/*fib*, Paper 3.1.2.

Fehling, E., Bunje, K., and Leutbecher, T., "Design Relevant Properties of Hardened Ultra High Performance Concrete," *Proceedings of the International Symposium on Ultra High Performance Concrete*, Ed., Schmidt, M., Fehling, E., and GeisenhanslLike, C., Kassel University Press, Kassel, Germany, 2004, pp. 327–338.

Fehling, E., Leutbecher, T., and StLirwald, S., "Structural Behavior of Ultra High Performance Concrete Reinforced With Steel Fibers and Rebars," *Proceedings of the Third International fib Congress and Exhibition Incorporating the PCI Annual Convention and National Bridge Conference*, Washington, DC, May 29–June 2, 2010, Compact Disc, Paper 25.

Fehling, E., Lorenz, P., and Leutbecher, T., "Experimental Investigations on Anchorage of Rebars in UHPC," *Proceedings of Hipermat 2012 3rd International Symposium on UHPC and Nanotechnology for High Performance Construction Materials*, Ed., Schmidt, M., Fehling, E., Glotzbach, C., Fröhlich, S., and Piotrowski, S., Kassel University Press, Kassel, Germany, 2012, pp. 533–540.

Felicetti, R. et al., "Mechanical Behaviour of HPC and UHPC in Direct Tension at High Temperature and After Cooling," *Proceedings of the Fifth RILEM Symposium on Fibre-Reinforced Concretes (FRC)BEFIB' 2000*, Publication pro015, Lyon, France, Ed., Rossi, P. and Chanvillard, G., 2000, pp. 749–758.

Ferrier, E. et al., "Mechanical Behaviour of Ultra High-Performance Fibrous-Concrete Beams Reinforced by Internal FRP Bars," *Proceedings of Hipermat 2012 3rd International Symposium on UHPC and Nanotechnology for High Performance Construction Materials*, Ed., Schmidt, M., Fehling, E., Glotzbach, C., Fröhlich, S., and Piotrowski, S., Kassel University Press, Kassel, Germany, 2012, pp. 619–630.

Fidjestol, P., Thorsteinsen, R. J., and Svennevig, P., "Making UHPC with Local Materials—The Way Forward," *Proceedings of Hipermat 2012 3rd International Symposium on*

UHPC and Nanotechnology for High Performance Construction Materials, Ed., Schmidt, M., Fehling, E., Glotzbach, C., Fröhlich, S., and Piotrowski, S., Kassel University Press, Kassel, Germany, 2012, pp. 207–214.

Fitik, B., Niedermeier, R., and Zilch, K., "Fatigue Behaviour of Ultra-High Performance Concrete Under Cyclic Stress Reversal Loading," *Proceedings of the Third International fib Congress and Exhibition Incorporating the PCI Annual Convention and National Bridge Conference*, Washington, DC, May 29–June 2, 2010, Compact Disc, Paper 20.

Fitik, B., Niedermeier, R., and Zilch, K., "Fatigue Behaviour of Ultra High-Performance Concrete Under Cyclic Stress Reversal Loading," *Proceedings of the Second International Symposium on Ultra High Performance Concrete*, Ed., Fehling, E., Schmidt, M., and Stürwald. S., Kassel University Press, Kassel, Germany, 2008, pp. 529–536.

Flietstra, J.C. et al., "Creep Behavior of UHPC under Compressive Loading with Varying Curing Regimes," *Proceedings of Hipermat 2012 3rd International Symposium on UHPC and Nanotechnology for High Performance Construction Materials*, Ed., Schmidt, M., Fehling, E., Glotzbach, C., Fröhlich, S., and Piotrowski, S., Kassel University Press, Kassel, Germany, 2012, pp. 333–340.

Francisco, P. et al., "Creep and Shrinkage Prediction for a Heat-Treated Ultra High Performance Fibre-Reinforced Concrete," *Proceedings of Hipermat 2012 3rd International Symposium on UHPC and Nanotechnology for High Performance Construction Materials*, Ed., Schmidt, M., Fehling, E., Glotzbach, C., Fröhlich, S., and Piotrowski, S., Kassel University Press, Kassel, Germany, 2012, pp. 325–331.

Francisco, P. et al., "Prediction of Ultra High Performance Concrete Creep and Shrinkage—The Role of Heat Treatment," *Proceedings of the Third International fib Congress and Exhibition Incorporating the PCI Annual Convention and National Bridge Conference*, Washington, DC, May 29–June 2, 2010, Compact Disc, Paper 219.

Francisco, P. et al., "Ultra High Performance Concrete for Prestressed Elements—Interest of Creep Prediction," *Proceedings of the International Workshop on Ultra High Performance Fibre Reinforced Concrete—Designing and Building with UHPFRC: State of the Art Development*, 17-19 November 2009, Marseille, France, AFGC/*fib*, Paper 5.2.1.

Freisinger, S. et al., "Structural and Semi-Structural Adhesive Bonding of UHPC by Modifying the Surface and Close to Surface Layers," *Proceedings of the Second International Symposium on Ultra High Performance Concrete*, Ed., Fehling, E., Schmidt, M., and Stürwald. S., Kassel University Press, Kassel, Germany, 2008, pp. 275–282.

Frettlöhr, B. et al., "Tests on the Flexural Tensile Strength of a UHPFRC Subjected to Cycling and Reversed Loading," *Proceedings of Hipermat 2012 3rd International Symposium on UHPC and Nanotechnology for High Performance Construction Materials*, Ed., Schmidt, M., Fehling, E., Glotzbach, C., Fröhlich, S., and Piotrowski, S., Kassel University Press, Kassel, Germany, 2012, pp. 427–434.

Freytag, B. et al., "Buckling Behaviour of UHPFRC-Panels under Compression," *Proceedings of Hipermat 2012 3rd International Symposium on UHPC and Nanotechnology for High Performance Construction Materials*, Ed., Schmidt, M.,

Fehling, E., Glotzbach, C., Fröhlich, S., and Piotrowski, S., Kassel University Press, Kassel, Germany, 2012, pp. 765–772.

Freytag, B. et al., "The Use of UHPC in Composites—Ideas and Realisations," *Proceedings of the International Symposium on Ultra High Performance Concrete*, Ed., Schmidt, M., Fehling, E., and Geisenhanslüke, C., Kassel University Press, Kassel, Germany, 2004, pp. 559–571.

Freytag, B. et al., "WILD-Bridge Scientific Preparation for Smooth Realisation," *Proceedings of Hipermat 2012 3rd International Symposium on UHPC and Nanotechnology for High Performance Construction Materials*, Ed., Schmidt, M., Fehling, E., Glotzbach, C., Fröhlich, S., and Piotrowski, S., Kassel University Press, Kassel, Germany, 2012, pp. 881–888.

Fröhlich, S. and Schmidt, M., "Influences on Repeatability and Reproducibility of Testing Methods for Fresh UHPC," *Proceedings of Hipermat 2012 3rd International Symposium on UHPC and Nanotechnology for High Performance Construction Materials*, Ed., Schmidt, M., Fehling, E., Glotzbach, C., Fröhlich, S., and Piotrowski, S., Kassel University Press, Kassel, Germany, 2012, pp. 225–232.

Gallwoszus, J., Hegger, J., and Heinemeyer, S., "Design Models for Composite Beams With Puzzle Strip Shear Connector and UHPC," *Proceedings of Hipermat 2012 3rd International Symposium on UHPC and Nanotechnology for High Performance Construction Materials*, Ed., Schmidt, M., Fehling, E., Glotzbach, C., Fröhlich, S., and Piotrowski, S., Kassel University Press, Kassel, Germany, 2012, pp. 647–654.

Gao, R., Stroeven, P., and Hendriks, C.F., "Mechanical Properties of Reactive Powder Concrete Beams," *Seventh International Symposium on the Utilization of High-Strength/High-Performance Concrete*, Vol. II, Publication No. SP-228, Ed., Russell, H.G., American Concrete Institute, Farmington Hills, MI, 2005, pp. 1,237–1,252.

Garas, V.Y., "Multi-Scale Investigation of Tensile Creep of Ultra-High Performance Concrete for Bridge Applications," PhD Dissertation, Georgia Institute of Technology, Atlanta, GA, 2009.

Garas, V.Y., Kahn, L.F., and Kurtis, K.E., "Short-Term Tensile Creep and Shrinkage of Ultra-High Performance Concrete," *Cement and Concrete Composites*, Vol. 31, No. 3, March 2009, pp. 47–152.

Garcia, H.M., "Analysis of an Ultra-High Performance Concrete Two-Way Ribbed Bridge Deck Slab," FHWA, U.S. Department of Transportation, Report No. FHWA-HRT-07-056. Available through National Technical Information Service at http://www.ntis.gov with Accession No. PB2007112112 [Cited November 23, 2011].

Gilbert, R.I., Gowripalan, N., and Cavill, B., "On the Design of Precast, Prestressed Reactive Powder Concrete (Ductal) Girders," *Proceedings of the Fourth Austroads Bridge Conference*, 29 November–1 December 2000, Adelaide, South Australia, Vol. 3, pp. 313–324.

Gilliland, S.K., "Reactive Powder Concrete (RPC), a New Material for Prestressed Concrete Bridge Girders," *Proceedings of the Fourteenth Structures Congress, Building an International*

Community of Structural Engineers, April 15–18, 1996, Chicago, IL, Ed., Ghosh, S.K. and Mohammadi, J., ASCE, Vol. 1, pp. 125–132.

Gowripalan, N. and Gilbert, R.I., "Design Guidelines for Ductal Prestressed Concrete Beams," The University of New South Wales, Sydney, Australia, 2000.

Gowripalan, N. et al., "Reactive Powder Concrete (RPC) for Precast Structural Concrete—Research and Development in Australia," *Proceedings of the Twenty-first Biennial Conference of the Concrete Institute of Australia—Concrete in the Third Millennium*, 17–19 July 2003, Brisbane, Australia, pp. 99–108.

Graybeal, B.A. and Hartmann, J.L., "Construction of an Optimized UHPC Vehicle Bridge," *Seventh International Symposium on the Utilization of High-Strength/High-Performance Concrete*, Vol. II, Publication No. SP-228, Ed., Russell, H.G., American Concrete Institute, Farmington Hills, MI, 2005, pp. 1,109–1,118.

Graybeal, B.A. and Hartmann, J.L., "Strength and Durability of Ultra-High Performance Concrete," *Proceedings of the 3rd International Symposium on High Performance Concrete/PCI National Bridge Conference*, October 19–22, 2003, Orlando, FL, Compact Disc, Paper 47.

Graybeal, B.A. and Hartmann, J.L., "Ultra-High Performance Concrete for Prestressed Concrete Bridge Girders," *Proceedings of the Concrete Bridge Conference*, May 17–18, 2004, Charlotte, NC.

Graybeal, B.A. and Hartmann, J.L., "Ultra-High Performance Concrete Material Properties," Transportation Research Board Annual Meeting, 2003, Washington, DC, Compact Disc.

Graybeal, B.A. and Swenty, M., "UHPFRC for Prefabricated Bridge Component Connections," *Proceedings of Hipermat 2012 3rd International Symposium on UHPC and Nanotechnology for High Performance Construction Materials*, Ed., Schmidt, M., Fehling, E., Glotzbach, C., Fröhlich, S., and Piotrowski, S., Kassel University Press, Kassel, Germany, 2012, pp. 663–668.

Graybeal, B.A. et al., "Direct and Flexural Test Methods for Determination of the Tensile Stress-Strain Response of UHPFRC," *Proceedings of Hipermat 2012 3rd International Symposium on UHPC and Nanotechnology for High Performance Construction Materials*, Ed., Schmidt, M., Fehling, E., Glotzbach, C., Fröhlich, S., and Piotrowski, S., Kassel University Press, Kassel, Germany, 2012, pp. 395–402.

Graybeal, B.A., "Design, Fabrication, and Testing of a 2nd Generation Ultra-High Performance Concrete Pi-Girder," *Proceedings of the Third International fib Congress and Exhibition Incorporating the PCI Annual Convention and National Bridge Conference*, Washington, DC, May 29–June 2, 2010, Compact Disc, Paper 94.

Graybeal, B.A., "Behavior of Field-Cast Ultra-High Performance Concrete Bridge Deck Connections Under Cyclic and Static Structural Loading," FHWA, U.S. Department of Transportation, Report No. FHWA-HRT-11-023, National Technical Information Service Accession No. PB2011-101995, 2010.

Graybeal, B.A., "Characterization of the Behavior of Ultra-High Performance Concrete," PhD Thesis, University of Maryland, 2005.

Graybeal, B.A., "Compressive Behavior of Ultra-High-Performance Fiber-Reinforced Concrete," *ACI Materials Journal*, Vol. 104, No. 2, March–April 2007, pp. 146–152.

Graybeal, B.A., "Fabrication of an Optimized UHPC Bridge," *Proceedings of the PCI National Bridge Conference*, October 17–20, 2004, Atlanta, GA, Compact Disc, Paper 16.

Graybeal, B.A., "Flexural Behavior of an Ultrahigh-Performance Concrete I-Girder," *Journal of Bridge Engineering*, Vol. 13, No. 6, 2008, pp. 602–610.

Graybeal, B.A., "Practical Means for Determination of the Tensile Behavior of Ultra-High Performance Concrete," *Journal of ASTM International*, Vol. 3, No. 8, September 2006.

Graybeal, B.A., "Simultaneous Structural and Environmental Loading of an Ultra-High Performance Concrete Component," FHWA, U.S. Department of Transportation, Report No. FHWA-HRT-10-054. Available through National Technical Information Service at http://www.ntis.gov with Accession No. PB2010110331 [Cited November 23, 2011].

Graybeal, B.A., "Structural Behavior of a 2nd Generation Ultra-High Performance Concrete Pi-Girder," FHWA, U.S. Department of Transportation, Report No. FHWA-HRT-10-026, 2009, 114 pp. Available through National Technical Information Service at http://www.ntis.gov with Accession No. PB2009115496 [Cited November 23, 2011].

Graybeal, B.A., "Structural Behavior of a Prototype Ultra-High Performance Concrete Pi-Girder," FHWA, U.S. Department of Transportation, Report No. FHWA-HRT-10-027. Available through National Technical Information Service at http://www.ntis.gov with Accession No. PB2009115495 [Cited November 23, 2011].

Graybeal, B.A., "Structural Behavior of Ultra-High Performance Concrete Prestressed I-Girders," FHWA, U.S. Department of Transportation, Report No. FHWA-HRT-06-115, 2006. Available through National Technical Information Service at http://www.ntis.gov with Accession No. PB2007104386 [Cited November 23, 2011].

Graybeal, B.A., "UHPC Making Strides," *Public Roads*, Vol. 72, No. 4, January/February 2009. Available at http://www.fhwa.dot.gov/publications/publicroads/09janfeb/03.cfm [Cited March 3, 2012].

Graybeal, B. and Baby, F., "Development of a Direct Tension Test Method for UHPFRC," *ACI Materials Journal*, Vol. 110, No. 2, March–April 2013.

Graybeal, B. and Davis, M., "Cylinder or Cube: Strength Testing of 80 to 200 MPa (11.6 to 29 ksi) Ultra-High-Performance Fiber-Reinforced Concrete," *ACI Materials Journal*, Vol. 105, No. 6, November–December 2008, pp. 603–609.

Graybeal, B. and Hartmann, J., "Experimental Testing of UHPC Optimized Bridge Girders: Early Results," *Proceedings of the PCI 51st Annual Convention/Exhibition and the PCI National Bridge Conference,* Palm Springs, CA, October 16–19, 2005, Compact Disc.

Graybeal, B. and Stone, B., "Compression Response of a Rapid-Strengthening Ultra-High Performance Concrete Formulation," FHWA, U.S. Department of Transportation, Report No. FHWA-HRT-12-065, National Technical Information Service Accession No. PB2012-112545, 2012.

Graybeal, B. and Tanesi, J., "Durability of an Ultrahigh-Performance Concrete," *ASCE Journal of Materials in Civil Engineering*, Vol. 19, No. 10, October 2007, pp. 848–854.

Graybeal, B., "Analysis of an Ultra-High Performance Concrete Two-Way Ribbed Bridge Deck Slab," *TechBrief*, FHWA-HRT-07-055, Federal Highway Administration, McLean, VA, 2007.

Graybeal, B., *Construction of Field-Cast Ultra-High Performance Concrete Connections*, *TechNote*, FHWA-HRT-12-038, Federal Highway Administration, McLean, VA, 2012.

Graybeal, B., "Fabrication of an Optimized UHPC Bridge," *Proceedings of the PCI National Bridge Conference*, October 17–20, 2004, Atlanta, GA, Compact Disc, Paper 16.

Graybeal, B., "Field-Cast UHPC Connections for Modular Bridge Deck Elements," *HPC Bridge Views*, Issue No. 66, March/April 2011. Available at http://www.hpcbridgeviews.org [Cited November 23, 2011].

Graybeal, B., *Field-Cast UHPC Connections for Modular Bridge Deck Elements, TechBrief*, FHWA-HRT-11-022, Federal Highway Administration, McLean, VA, 2010.

Graybeal, B., "Finite Element Analysis of UHPC: Structural Performance of an AASHTO Type II Girder and a 2nd-Generation Pi-Girder," *TechBrief*, FHWA-HRT-10-079, Federal Highway Administration, McLean, VA, 2010.

Graybeal, B., "Material Property Characterization of Ultra-High Performance Concrete," FHWA, U.S. Department of Transportation, Report No. FHWA-HRT-06-103, McLean, VA, 2006.

Graybeal, B., "Simultaneous Structural and Environmental Loading of a UHPC Component," *TechBrief*, FHWA-HRT-10-055, Federal Highway Administration, McLean, VA, 2010.

Graybeal, B., *Structural Behavior of a 2nd-Generation UHPC Pi-Girder, TechBrief*, FHWA-HRT09-069, 2009 Federal Highway Administration, McLean, VA,.

Graybeal, B., *Structural Behavior of a Prototype UHPC Pi-Girder, TechBrief*, FHWA-HRT-09- 068, Federal Highway Administration, McLean, VA, 2009.

Graybeal, B., "UHPC for PBES Connections." Available at http://www.abc.fiu.edu/wp-content/uploads/2011/10/Graybeal-UHPC-20110929Webinar-distributeipdf [Cited March 12, 2012].

Graybeal, B., "UHPC for Structural Connections," ACI Convention Presentation: UHPC-Experience and Developments, October 22, 2012. Available at http://www.concrete.org /Convention/fall-Convention/PresentationDetail.asp?EventId=ZSESS14 [Cited December 14, 2012].

Graybeal, B., "UHPC in the U.S. Highway Infrastructure," *Proceedings of the International Workshop on Ultra High Performance Fibre Reinforced Concrete—Designing and Building with UHPFRC: State of the Art Development*, 17–19 November 2009, Marseille, France, AFGC/*fib*, Paper 3.2.2.

Graybeal, B., "UHPC in the U.S. Highway Transportation System," *Proceedings of the Second International Symposium on Ultra High Performance Concrete*, Ed., Fehling, E., Schmidt, M., and Stürwald. S., Kassel University Press, Kassel, Germany, 2008, pp. 11–17.

Graybeal, B., "Ultra-High Performance Concrete Composite Connections for Precast Concrete Bridge Decks," FHWA, U.S. Department of Transportation, Report No. FHWA-HRT-12-041. Available through National Technical Information Service at http://www.ntis.gov with Accession No. PB2012107569 [Cited April 13, 2012]. To be published.

Graybeal, B., *Ultra-High Performance Concrete, TechNote*, FHWA-HRT-11-038, Federal Highway Administration, McLean, VA, 2011.

Graybeal, B., "Using UHPC Connections for Precast Concrete Bridge Decks," *ASPIRE*, Summer 2011, p. 50. Available at http://www.aspirebridge.org [Cited November 23, 2011].

Graybeal, B., Hartmann, J., and Perry, V., "Ultra-High Performance Concrete for Highway Bridges," *Proceedings of the fib Symposium: Concrete Structures: The Challenge of Creativity*, 26–28 April 2004, Avignon, France, AFGC.

Graybeal, B., Perry, V., and Royce, M., "UHPC Ultra-High Performance Concrete," NHI Innovations Webinar, November 18, 2010. Available at https://connectdot.connectsolutions.com/n134083201011 [Cited April 3, 2012].

Gröger, J., Viet Tue, N., and Wille, K., "Bending Behaviour and Variation of Flexural Parameters of UHPFRC," *Proceedings of Hipermat 2012 3rd International Symposium on UHPC and Nanotechnology for High Performance Construction Materials*, Ed., Schmidt, M., Fehling, E., Glotzbach, C., Fröhlich, S., and Piotrowski, S., Kassel University Press, Kassel, Germany, 2012, pp. 419–426.

Grünberg, J. et al., "Multi-Axial and Fatigue Behaviour of Ultra-High-Performance Concrete," *Proceedings of the Second International Symposium on Ultra High Performance Concrete*, Ed., Fehling, E., Schmidt, M., and Stürwald. S., Kassel University Press, Kassel, Germany, 2008, pp. 485–492.

Grünewald, S. and Van Loenhout, P., "Applications of UHPC: Outstanding Projects from Hurks Beton," *Betonwerk und Fertigteil Technick Journal*, Vol. 74, No. 11, 2008, pp. 26–34.

GrLinewald, S., Weyns, R., and Dekkers, J., "Experience with Prefabricated UHPFRC in the Netherlands," *Proceedings of the International Workshop on Ultra High Performance Fibre Reinforced Concrete—Designing and Building with UHPFRC: State of the Art Development*, 17–19 November 2009, Marseille, France, AFGC/*fib*, Paper 4.2.1.

GLivensoy, G. et al., "Mechanical Behavior of High Performance Steel Fiber Reinforced Cementitious Composites Under Cyclic Loading Condition," *Proceedings of the International Symposium on Ultra High Performance Concrete*, Ed., Schmidt, M., Fehling, E., and GeisenhanslLike, C., Kassel University Press, Kassel, Germany, 2004, pp. 649–660.

Habel, K. and Gauvreau, P., "Behavior of Reinforced and Posttensioned Concrete Members With a UHPFRC Overlay Under Impact Loading," *Journal of Structural Engineering*, Vol. 135, No. 3, 2009, pp. 292–300.

Habel, K. and Gauvreau, P., "Response of Ultra-High Performance Fiber Reinforced Concrete (UHPFRC) to Impact and Static Loading," *Cement and Concrete Composites*, Vol. 30, No. 10, 2008, pp. 938–946.

Habel, K. et al., "Development of the Mechanical Properties of an Ultra-High Performance Fiber Reinforced Concrete (UHPFRC)," *Cement and Concrete Research*, Vol. 36, No. 7, July 2006, pp. 1,362–1,370.

Habel, K. et al., "Ultra-High Performance Fibre Reinforced Concrete Mix Design in Central Canada," *Canadian Journal of Civil Engineering*, Vol. 35, No. 2, February 2008, pp. 217–224.

Habel, K., Denarié, E., and BrLihwiler, E., "Experimental Investigation of Composite Ultra-HighPerformance Fiber-Reinforced Concrete and Conventional Concrete Members," *ACI Structural Journal*, Vol. 104, No. 1, January–February 2007, pp. 93–101.

Habel, K., Denarié, E., and BrLihwiler, E., "Structural Elements Combining Ultra-High Performance Fibre Reinforced Concrete (UHPFRC) and Concrete: An Innovative Solution for Rehabilitation and Modification of Existing Concrete Structures," *Beton und Stahlbetonbau*, Vol. 100, No. 2, February 2005, pp. 124–131.

Habel, K., Denarié, E., and BrLihwiler, E., "Structural Response of Composite 'UHPFRC-Concrete' Members Under Bending," *Proceedings of the International Symposium on*

Ultra High Performance Concrete, Ed., Schmidt, M., Fehling, E., and GeisenhanslLike, C., Kassel University Press, Kassel, Germany, 2004, pp. 389–399.

Habel, K., Denarié, E., and BrLihwiler, E., "Structural Response of Elements Combining Ultrahigh-Performance Fiber-Reinforced Concretes and Reinforced Concrete," *Journal of Structural Engineering*, Vol. 132, No. 11, 2006, pp. 1,793–1,800.

Habel, K., Denarié, E., and BrLihwiler, E., "Time Dependent Behavior of Elements Combining Ultra-High Performance Fiber Reinforced Concretes (UHPFRC) and Reinforced Concrete," *Materials and Structures*, Vol. 39, No. 5, 2006, pp. 557–569.

Hackman, L.E., Farrell, M.B., and Dunham, O.O., "Ultra High Performance Reinforced Concrete," *Fiber Reinforced Concrete Developments and Innovations*, Publication No. SP-142, Ed., Daniel, J.I. and Shah, S.P., American Concrete Institute, Farmington Hills, MI, 1994, pp. 235–248.

Hajar, Z. et al., "Construction of the First Road Bridges Made of Ultra-High-Performance Concrete," *Proceedings of the 3rd International Symposium on High Performance Concrete/PCI National Bridge Conference*, October 19–22, 2003, Orlando, FL, Compact Disc, Paper 76.

Hajar, Z. et al., "Design and Construction of the World First Ultra-High Performance Concrete Road Bridges," *Proceedings of the International Symposium on Ultra High Performance Concrete*, Ed., Schmidt, M., Fehling, E., and GeisenhanslLike, C., Kassel University Press, Kassel, Germany, 2004, pp. 39–48.

Hajar, Z. et al., "La Poutre ITE®, une Alternative Economique et Durable aux Poutrelles Enrobees," (The ITE UHPC Girder, an Economical and Durable Alternative to Encased Steel Girders), *Proceedings of the International Workshop on Ultra High Performance Fibre Reinforced Concrete—Designing and Building with UHPFRC: State of the Art Development*, 17–19 November 2009, Marseille, France, AFGC/*fib*, Paper 3.2.3, 14 pp.

Hajar, Z. et al., "Ultra-High-Performance Concretes: First Recommendations and Examples of Applications," *Proceedings of the fib Symposium: Concrete Structures: The Challenge of Creativity*, 26–28 April 2004, Avignon, France, AFGC, pp. 242–243.

Hanoteau, B. et al., "Ductal: A New Material, the Bridge of St. Pierre La Cour, in the French Technology of Concrete," *Proceedings of the Second fib Congress*, June 5–8, 2006, Naples, Italy. Available on Compact Disc from http://www.fib-international.org/proceedings-of-previousconferences [Cited November 23, 2011].

Hansen, L. and Jensen, B. (1999). "A New Building System Using Joints of Ultra High-Strength Fibre Reinforced Concrete," *Innovation in Concrete Structures: Design and Construction: Proceedings of the International Conference on Creating with Concrete*, University of Dundee, Scotland, pp. 543–552.

Harris, D.K. and Roberts-Wollmann, C.L., "Characterization of Punching Shear Capacity of Thin Ultra-High Performance Concrete Slabs," *Proceedings of the Second International Symposium on Ultra High Performance Concrete*, Ed., Fehling, E., Schmidt, M., and StLirwald. S., Kassel University Press, Kassel, Germany, 2008, pp. 727–734.

Harris, D.K. and Roberts-Wollmann, C.L., "Characterization of the Punching Shear Capacity of Thin Ultra-High Performance Concrete Slabs," Virginia Transportation Research Council, Report No. VTRC 05-CR26, Richmond, VA, 2005.

Harris, D., "Characterization of the Punching Shear Capacity of Thin UHPC Plates," Master's Thesis, Virginia Polytechnic Institute and State University, Blacksburg, VA, 2004.

Harris, D., Sarkar, J., and Ahlborn, T., "Interface Bond Characterization of UHPC Overlays," *Transportation Research Board 90th Annual Meeting Compendium of Papers*, January 23–27, 2011, Washington, DC, Compact Disc, Session 239, Paper 11-1481.

Hartmann, J. and Graybeal, B., "Testing of Ultra-High-Performance Concrete Girders," *PCI Journal*, Vol. 47, No. 1, January/February 2002, pp. 148–149.

Hartmann, J. and Graybeal, B., "The FHWA Ultra-High-Performance Concrete Program," *Proceedings of the PCI National Bridge Conference*, October 6–9, 2002, Nashville, TN, Compact Disc, Paper 15, 1 pp. Abstract only.

Hassan, A. et al., "Evaluation of Bond Strength Between Ultra-High Performance Reactive Powder Composite Materials and Fiber-Reinforced Concrete by Slant Shear Test," *Seventh CANMET/ACI International Conference on Recent Advances in Concrete Technology,* Publication No. SP-222, Ed., Malhotra, V.M., American Concrete Institute, Farmington Hills, MI, 2004, pp. 215–230.

Hecht, M., "Practical Use of Fibre-Reinforced UHPC in Construction—Production of Precast Elements for Wild-Brücke in Völkermarkt," *Proceedings of Hipermat 2012 3rd International Symposium on UHPC and Nanotechnology for High Performance Construction Materials*, Ed., Schmidt, M., Fehling, E., Glotzbach, C., Fröhlich, S., and Piotrowski, S., Kassel University Press, Kassel, Germany, 2012, pp. 889–896.

Hegger, J. and Bertram, G., "Anchorage Behavior of Pretensioned Strands in Steel Fiber Reinforced UHPC," *Proceedings of the Second International Symposium on Ultra High Performance Concrete*, Ed., Fehling, E., Schmidt, M., and Stürwald. S., Kassel University Press, Kassel, Germany, 2008, pp. 537–544.

Hegger, J. and Bertram, G., "Shear Carrying Capacity of Steel Fiber Reinforced UHPC," *Proceedings of the Second International Symposium on Ultra High Performance Concrete*, Ed., Fehling, E., Schmidt, M., and Stürwald. S., Kassel University Press, Kassel, Germany, 2008, pp. 513–520.

Hegger, J. and Rauscher, S., "UHPC in Composite Construction," *Proceedings of the Second International Symposium on Ultra High Performance Concrete*, Ed., Fehling, E., Schmidt, M., and Stürwald. S., Kassel University Press, Kassel, Germany, 2008, pp. 545–552.

Hegger, J. et al., "Connections of Precast UHPC Elements," *Proceedings of Hipermat 2012 3rd International Symposium on UHPC and Nanotechnology for High Performance Construction Materials*, Ed., Schmidt, M., Fehling, E., Glotzbach, C., Fröhlich, S., and Piotrowski, S., Kassel University Press, Kassel, Germany, 2012, pp. 655–662.

Hegger, J., Rauscher, S., and Gallwoszus, J., "Modern Hybrid Structures Made of UHPC and High Strength Steel," *Proceedings of the Third International fib Congress and Exhibition Incorporating the PCI Annual Convention and National Bridge Conference*, Washington, DC, May 29–June 2, 2010, Compact Disc, Paper 52.

Hegger, J., Rauscher, S., and Goralski, C., "Push-Out Tests on Headed Studs Embedded in UHPC," *Proceedings of the International Symposium on Ultra High Performance Concrete*, Ed., Schmidt, M., Fehling, E., and GeisenhanslLike, C., Kassel University Press, Kassel, Germany, 2004, pp. 425–434.

Hegger, J., Tuchlinski, D., and Komner, B., "Bond Anchorage Behavior and Shear Capacity of Ultra High Performance Concrete Beams," *Proceedings of the International Symposium on Ultra High Performance Concrete*, Ed., Schmidt, M., Fehling, E., and GeisenhanslLike, C., Kassel University Press, Kassel, Germany, 2004, pp. 351–360.

Heimann, J. and Schuler, G., "The Implementation of Full Depth UHPC Waffle Bridge Deck Panels Phase 1 Report," FHWA Highways for LIFE Technology Partnerships Program, DTFH61-09-G-00006, Federal Highway Administration, McLean, VA, 2010.

Heinz, D. and Ludwig, H.-M., "Heat Treatment and the Risk of DEF Delayed Ettringite Formation in UHPC," *Proceedings of the International Symposium on Ultra High Performance Concrete*, Ed., Schmidt, M., Fehling, E., and GeisenhanslLike, C., Kassel University Press, Kassel, Germany, 2004, pp. 717–730.

Heinz, D., Dehn, F., and Urbonas, L., "Fire Resistance of Ultra High Performance Concrete (UHPC)—Testing of Laboratory Samples and Columns Under Load," *Proceedings of the International Symposium on Ultra High Performance Concrete*, Ed., Schmidt, M., Fehling, E., and GeisenhanslLike, C., Kassel University Press, Kassel, Germany, 2004, pp. 703–715.

Heinz, D., Urbonas, L., and Gerlicher, T., "Effect of Heat Treatment Method on the Properties of UHPC," *Proceedings of Hipermat 2012 3rd International Symposium on UHPC and Nanotechnology for High Performance Construction Materials*, Ed., Schmidt, M., Fehling, E., Glotzbach, C., Fröhlich, S., and Piotrowski, S., Kassel University Press, Kassel, Germany, 2012, pp. 283–290.

Hoffmann, S. and Weiher, H., "Innovative Design of Bridge Bearings by the Use of UHPFRC," *Proceedings of Hipermat 2012 3rd International Symposium on UHPC and Nanotechnology for High Performance Construction Materials*, Ed., Schmidt, M., Fehling, E., Glotzbach, C., Fröhlich, S., and Piotrowski, S., Kassel University Press, Kassel, Germany, 2012, pp. 973–980.

Holschemacher, K. and Weiße, D., "Economic Mix Design Ultra High-Strength Concrete," *Seventh International Symposium on the Utilization of High-Strength/High-Performance Concrete*, Vol. II, Publication No. SP-228, Ed., Russell, H.G., American Concrete Institute, Farmington Hills, MI, 2005, pp. 1,133–1,144.

Holschemacher, K. et al., "Experimental Investigation on Ultra High-Strength Concrete Under Concentrated Loading," *Seventh International Symposium on the Utilization of HighStrength/High-Performance Concrete*, Vol. II, Publication No. SP-228, Ed., Russell, H.G., American Concrete Institute, Farmington Hills, MI, 2005, pp. 1,145–1,157.

Holschemacher, K. et al., "Ultra High Strength Concrete Under Concentrated Loading," *Proceedings of the International Symposium on Ultra High Performance Concrete*, Ed., Schmidt, M., Fehling, E., and Geisenhanslüke, C., Kassel University Press, Kassel, Germany, 2004, pp. 471–480.

Holschemacher, K., Weiße, D., and Klotz, S., "Bond of Reinforcement in Ultra High-Strength Concrete," *Seventh International Symposium on the Utilization of High-Strength/High-Performance Concrete*, Vol. I, Publication No. SP-228, Ed., Russell, H.G., American Concrete Institute, Farmington Hills, MI, 2005, pp. 513–528.

Holschemacher, K., Weiβe, D., and Klotz, S., "Bond of Reinforcement in Ultra High Strength Concrete," *Proceedings of the International Symposium on Ultra High Performance Concrete*, Ed., Schmidt, M., Fehling, E., and Geisenhanslüke, C., Kassel University Press, Kassel, Germany, 2004, pp. 375–387.

Hong, S.-G. et al., "Thermal Properties of Mortar With Ceramic Microspheres," *Proceedings of Hipermat 2012 3rd International Symposium on UHPC and Nanotechnology for High Performance Construction Materials*, Ed., Schmidt, M., Fehling, E., Glotzbach, C.,

Fröhlich, S., and Piotrowski, S., Kassel University Press, Kassel, Germany, 2012, pp. 557–564.

Horszczaruk, E., "Abrasion Resistance of High Strength Fibre-Reinforced Concrete," *Sixth RILEM Symposium on Fibre-Reinforced Concretes (FRC), BEFIB 2004*, Publication Pro039, Varenna, Italy, September 20–22, 2004, pp. 257–266.

Hossain, K., et al., "Bond Strength and Development Length of GFRP Bars in Ultra-High Performance Concrete," Final report, Ontario Ministry of Transportation Research Project, Department of Civil Engineering, Ryerson University, Toronto, Canada, 2011.

Hosser, D., Kampmeier, B., and Hollmann, D., "Behavior of Ultra High Performance Concrete (UHPC) in Case of Fire," *Proceedings of Hipermat 2012 3rd International Symposium on UHPC and Nanotechnology for High Performance Construction Materials*, Ed., Schmidt, M., Fehling, E., Glotzbach, C., Fröhlich, S., and Piotrowski, S., Kassel University Press, Kassel, Germany, 2012, pp. 573–582.

Ibuk, H. and Beckhaus, K., "Ultra High Performance Concrete for Drill Bits in Special Foundation Engineering," *Proceedings of Hipermat 2012 3rd International Symposium on UHPC and Nanotechnology for High Performance Construction Materials*, Ed., Schmidt, M., Fehling, E., Glotzbach, C., Fröhlich, S., and Piotrowski, S., Kassel University Press, Kassel, Germany, 2012, pp. 807–810.

Ichinomiya, T. et al., "Experimental Study on Mechanical Properties of Ultra-High-Strength Concrete with Low-Autogenous-Shrinkage," *Seventh International Symposium on the Utilization of High-Strength/High-Performance Concrete*, Vol. II, Publication No. SP-228, Ed., Russell, H.G., American Concrete Institute, Farmington Hills, MI, 2005, pp. 1,341–1,352.

Iowa Department of Transportation, "Special Provisions for Ultra High Performance Concrete," SP-090112a, Effective Date, February 15, 2011. Available at http://www.iowadot.gov/us6kegcreek/documents/SP-concrete.pdf [Cited May 29, 2012].

Jensen, B.C., "Applications of Steel-Fibre-Reinforced Ultra-High-Strength Concrete," *Structural Engineering International*, Vol. 9, No. 2, May 1999, pp. 143–146.

Joh, C. et al., "Punching Shear Strength Estimation of UHPC Slabs," *Proceedings of the Second International Symposium on Ultra High Performance Concrete*, Ed., Fehling, E., Schmidt, M., and Stürwald. S., Kassel University Press, Kassel, Germany, 2008, pp. 719–726.

Joh, C. et al., "Torsional Test of Ultra High Performance Fiber-Reinforced Concrete Square Members," *Proceedings of Hipermat 2012 3rd International Symposium on UHPC and Nanotechnology for High Performance Construction Materials*, Ed., Schmidt, M., Fehling, E., Glotzbach, C., Fröhlich, S., and Piotrowski, S., Kassel University Press, Kassel, Germany, 2012, pp. 509–516.

Joh, C., Hwang, H., and Kim, B., "Punching Shear and Flexural Strengths of Ultra High Performance Concrete Slabs," *High Performance Structures and Materials IV*, 13–15 May 2008, Algarve, Portugal, Ed., De Wilde W.P., and Brebbia, C.A., pp. 97–106.

Jovanovic, I., Paatsch, A., and Durukal, A., "Ductal®: A New Generation of Ultra High Performance Fiber Reinforced Concrete," *Proceedings of the 6th International Symposium on High Strength/High Performance Concrete*, Leipzig, Germany, June 2002, Ed., König, G., Dehn, F., and Faust, T., Vol. 2, pp. 1,089–1,095.

JSCE, *Recommendations for Design and Construction of High Performance Fiber Reinforced Cement Composites With Multiple Fine Cracks (HPFRCC)*, Japan Society of Civil Engineers, Concrete Engineering Series 82, March 2008.

JSCE, *Recommendations for Design and Construction of Ultra High Strength Fiber Reinforced Concrete Structures (Draft)*, Japan Society of Civil Engineers, JSCE Guidelines for Concrete No. 9. 2006.

Juhart, J. et al., "Adhesion of Fine-Grained HPC and UHPC to Steel and Glass," *Proceedings of Hipermat 2012 3rd International Symposium on UHPC and Nanotechnology for High Performance Construction Materials*, Ed., Schmidt, M., Fehling, E., Glotzbach, C., Fröhlich, S., and Piotrowski, S., Kassel University Press, Kassel, Germany, 2012, pp. 717–724.

Jungwirth, J. and Muttoni, A., "Structural Behavior of Tension Members in UHPC," *Proceedings of the International Symposium on Ultra High Performance Concrete*, Ed., Schmidt, M., Fehling, E., and Geisenhanslüke, C., Kassel University Press, Kassel, Germany, 2004, pp. 533–544.

Jungwirth, J. et al., "Utilization of UHPC in Composite Structures—Lightweight Composite Structures (LCS)," *Proceedings of the Second International Symposium on Ultra High Performance Concrete*, Ed., Fehling, E., Schmidt, M., and StLirwald. S., Kassel University Press, Kassel, Germany, 2008, pp. 887–894.

Kabele, P., "Multiscale Framework for Modeling of Fracture in High Performance Fiber Reinforced Cementitious Composites," *Engineering Fracture Mechanics*, Vol. 74, 2007, pp. 194–209.

Kakizaki, M., Effect of Mixing Method on Mechanical Properties and Pore Structure of Ultra High-Strength Concrete, Katri Report No. 90, Kajima Corporation, Tokyo, 1992.

Kamen, A. et al., "Thermo-Mechanical Response of UHPFRC at Early Age—Experimental Study and Numerical Simulation," *Cement and Concrete Research*, Vol. 38, No. 6, June 2008, pp. 822–831.

Kamen, A. et al., "UHPFRC Tensile Creep at Early Age," *Materials and Structures*, Vol. 42, No. 1, January 2009, pp. 113–122.

Kamen, A., "Time Dependent Behavior of Ultra High Performance Fibre Reinforced Concrete," *Proceedings of the Sixth International PhD Symposium in Civil Engineering*, 23–26 August 2006, Zurich, Switzerland, Ed., Vogel, T., pp. 72–73.

Kamen, A., Denarié, E., and BrLihwiler, E., "Mechanical Behavior of Ultra High Performance Fibre Reinforced Concretes (UHPFRC) at Early Age, and Under Restraint," *Proceedings of the International Conference on Creep, Shrinkage and Durability of Concrete and Concrete Structures (CONCREEP 7)*, 12–14 September 2005, Nantes, France, pp. 591–596.

Kamen, A., Denarié, E., and BrLihwiler, E., "Thermal Effects on Physico-Mechanical Properties of Ultra-High-Performance Fiber-Reinforced Concrete," *ACI Materials Journal*, Vol. 104, No. 4, July–August 2007, pp. 415–423.

Kanakubo, T., "Tensile Characteristics Evaluation Method for Ductile Fiber-Reinforced Cementitious Composites," *Journal of Advanced Concrete Technology*, February 2006, Vol. 4, No. 1, pp. 3–17.

Kaptijn, N. and Blom, J., "A New Bridge Deck for the Kaag Bridges, the First CRC (Compact Reinforced Composite) Application in Civil Infrastructure," *Proceedings of the International Symposium on Ultra High Performance Concrete*, Ed., Schmidt, M.,

Fehling, E., and GeisenhanslLike, C., Kassel University Press, Kassel, Germany, 2004, pp. 49–57.

Kasuga, A. et al., “Study of a Bridge with a New Structural System Using Ultra High Strength Fiber Reinforced Concrete,” *Proceedings of the Third International fib Congress and Exhibition Incorporating the PCI Annual Convention and National Bridge Conference*, Washington, DC, May 29–June 2, 2010, Compact Disc, Paper 141.

Kawabata, T., Suzuki, M., and Sato, R., “Effect of Autogenous Shrinkage Before Prestressing on Prestress Loss of Pretensioned Prestressed Concrete Beams,” *Proceedings of the Third International fib Congress and Exhibition Incorporating the PCI Annual Convention and National Bridge Conference*, Washington, DC, May 29–June 2, 2010, Compact Disc, Paper 657.

Kazemi, S. and Lubell, A., “Influence of Specimen Size and Fiber Content on Mechanical Properties of Ultra-High-Performance Fiber-Reinforced Concrete,” *ACI Materials Journal*, Vol. 109, No. 6, November–December 2012, pp. 675–684.

Keierleber, B. et al., “FHWA, Iowa Optimize Pi Girder,” *ASPIRE*, Winter 2010, pp. 24–26. Available at http://www.aspirebridge.org [Cited November 23, 2011].

Keierleber, B. et al., “Design of Buchanan County, Iowa, Bridge, Using Ultra High-Performance Concrete and PI-Beam Cross Section,” *Proceedings of the PCI National Bridge Conference*, October 4–7, 2008, Orlando, FL, Compact Disc, Paper 27.

Khoury, G. and Algar, S., “Advanced Mechanical Characterisation of HPC and UHPC Concretes at High Temperatures,” Material Properties and Design; Durable Reinforced Concrete Structures, Ed., Schwesinger, P. and Wittmann, F.H., Weimar, Germany, Aedificatio Verlag, 1998, pp. 267–282.

Kim, B.-S. et al., “R&D Activities and Application of Ultra High Performance Concrete to Cable Stayed Bridges,” *Proceedings of Hipermat 2012 3rd International Symposium on UHPC and Nanotechnology for High Performance Construction Materials*, Ed., Schmidt, M., Fehling, E., Glotzbach, C., Fröhlich, S., and Piotrowski, S., Kassel University Press, Kassel, Germany, 2012, pp. 865–872.

Kim, J.K. et al., “Ductility of Ultra High-Strength Concrete Tied Columns Under Combined Axial Load and Lateral Force,” *Key Engineering Materials*, Vol. 348-349, 2007, pp. 609–612.

Kim, S. et al., “Shrinkage Behavior of Ultra High Performance Concrete at the Manufacturing Stage,” *Proceedings of Hipermat 2012 3rd International Symposium on UHPC and Nanotechnology for High Performance Construction Materials*, Ed., Schmidt, M., Fehling, E., Glotzbach, C., Fröhlich, S., and Piotrowski, S., Kassel University Press, Kassel, Germany, 2012, pp. 317–324.

Kim, S.W. et al., “Effect of Filling Method on Fibre Orientation and Dispersion and Mechanical Properties of UHPC,” *Proceedings of the Second International Symposium on Ultra High Performance Concrete*, Ed., Fehling, E., Schmidt, M., and Stürwald. S., Kassel University Press, Kassel, Germany, 2008, pp. 185–192.

Kimura, H. et al., “Seismic Behavior of 200 MPa Ultra-High-Strength Steel-Fiber Reinforced Concrete Columns Under Varying Axial Load,” *Journal of Advanced Concrete Technology*, Vol. 5, No. 2, 2007, pp. 193–200.

Kimura, H. et al., “Structural Design of 80-Story RC High-Rise Building Using 200 MPa UltraHigh-Strength Concrete,” *Journal of Advanced Concrete Technology*, Vol. 5, No. 2, 2007, pp. 181–191.

Kleymann, M., Girgis, A.M., and Tadros, M. K., "Developing User-Friendly and Cost-Effective Ultra-High Performance Concrete," *Proceedings of the NCBC Concrete Bridge Conference*, Reno, Nevada, May 7–10, 2006.

Klinghoffer, O. and Aarup, B., "Effect of Microcracks on Durability of Ultra High Strength Concrete," *Proceedings of the Fourth International Symposium on Corrosion of Reinforcement in Concrete Construction*, 1–4 July 1996, Robison College, Cambridge, pp. 611–619.

Kohlmeyer, C. et al., "Investigations on Embedded Shear Connectors for Lightweight Composite Structures," *Proceedings of the Third International fib Congress and Exhibition Incorporating the PCI Annual Convention and National Bridge Conference*, Washington, DC, May 29–June 2, 2010, Compact Disc, Paper 429.

Kollmorgen, G.A., "Impact of Age and Size on the Mechanical Behavior of an Ultra High Performance Concrete," Master's Thesis, Michigan Technological University, 2004.

Krelaus, R., Freisinger, S., and Schmidt, M., "Adhesive Bonding of UHPC Structural Members at the Gaertnerplatz Bridge in Kassel," *Proceedings of the Second International Symposium on Ultra High Performance Concrete*, Ed., Fehling, E., Schmidt, M., and StLirwald. S., Kassel University Press, Kassel, Germany, 2008, pp. 597–604.

Lachemi, M. et al., "Monitoring of the World's First Reactive Powder Concrete Bridge," *Proceedings of the Fifth International Conference on Short and Medium Span Bridges*, 13–16 July 1998, Calgary, Alberta, Canada.

Lallemant-Gamboa, I. et al., "Formulations, Characterizations and Applications of Ultra High-Performance Concrete," *Seventh International Symposium on the Utilization of HighStrength/High-Performance Concrete*, Vol. II, Publication No. SP-228, Ed., Russell, H.G., American Concrete Institute, Farmington Hills, MI, 2005, pp. 1,221–1,235.

Lappa, E.S., Braam, C.R., and Walraven, J.C., "Static and Fatigue Bending Tests of UHPC," *Proceedings of the International Symposium on Ultra High Performance Concrete*, Ed., Schmidt, M., Fehling, E., and GeisenhanslLike, C., Kassel University Press, Kassel, Germany, 2004, pp. 449–458.

Le, T.E. et al., "Structural Behaviour of a UHPFRC Flag Pavement," *Proceedings of the Second International Symposium on Ultra High Performance Concrete*, Ed., Fehling, E., Schmidt, M., and StLirwald. S., Kassel University Press, Kassel, Germany, 2008, pp. 663–670.

Lee, Y., Kang, S.T., and Kim, J.K., "Pullout Behavior of Inclined Steel Fiber in an Ultra-High Strength Cementitious Matrix," *Construction and Building Materials*, Vol. 24, No. 10, October 2010, pp. 2,030–2,041.

Lei, V.Y., "Construction of Malaysia's First 50-m-Long UHPdC Composite Road Bridge," *Concrete Technology*, Vol. 4, 2011, pp. 122–127.

Leonhardt, S., Lowke, D., and Gehlen, G., "Effect of Fibres on Impact Resistance of Ultra High Performance Concrete," *Proceedings of Hipermat 2012 3rd International Symposium on UHPC and Nanotechnology for High Performance Construction Materials*, Ed., Schmidt, M., Fehling, E., Glotzbach, C., Fröhlich, S., and Piotrowski, S., Kassel University Press, Kassel, Germany, 2012, pp. 811–817.

Leutbecher, T. and Fehling E., "Tensile Behavior of Ultra-High-Performance Concrete Reinforced With Reinforcing Bars and Fibers: Minimizing Fiber Content," *ACI Materials Journal*, Vol. 109, No. 2, March–April 2012, pp. 253–263.

Leutbecher, T. and Fehling, E., "Crack Formation and Tensile Behaviour of UHPC Reinforced With a Combination of Rebars and Fibres," *Proceedings of the Second International Symposium on Ultra High Performance Concrete*, Ed., Fehling, E., Schmidt, M., and StLirwald. S., Kassel University Press, Kassel, Germany, 2008, pp. 497–504.

Leutbecher, T. and Fehling, E., "Structural Behaviour of UHPC Under Tensile Stress and Biaxial Loading," *Proceedings of the International Symposium on Ultra High Performance Concrete*, Ed., Schmidt, M., Fehling, E., and GeisenhanslLike, C., Kassel University Press, Kassel, Germany, 2004, pp. 435–446.

Li, Q. and Xu, S., "Influence of Reinforcement Ratio on Performances of RUHTCC Flexural Members," *Proceedings of the Third International fib Congress and Exhibition Incorporating the PCI Annual Convention and National Bridge Conference*, Washington, DC, May 29–June 2, 2010, Compact Disc, Paper 175.

Li, Z. et al., "Uniaxial Tensile Behavior of Concrete Reinforced With Randomly Distributed Short Fibers," *ACI Materials Journal*, Vol. 95, No. 5, September–October 1998, pp. 564-574.

Lichtenfels, A., "Ultra-High Performance Fibre Reinforced Concrete for Shells," *Proceedings of the International Symposium on Ultra High Performance Concrete*, Ed., Schmidt, M., Fehling, E., and GeisenhanslLike, C., Kassel University Press, Kassel, Germany, 2004, pp. 827–838.

Link, M. and Weiland, M., "Structural Health Monitoring of the Gaertnerplatz Bridge Over the Fulda River in Kassel Based on Vibration Test Data and Stochastic Model Updating," *Proceedings of Hipermat 2012 3rd International Symposium on UHPC and Nanotechnology for High Performance Construction Materials*, Ed., Schmidt, M., Fehling, E., Glotzbach, C., Fröhlich, S., and Piotrowski, S., Kassel University Press, Kassel, Germany, 2012, pp. 945–956.

Link, M., Weiland, M., and Hahn, T., "Structural Health Monitoring of the Gaertnerplatz Bridge Over the Fulda River in Kassel," *Proceedings of the Second International Symposium on Ultra High Performance Concrete*, Ed., Fehling, E., Schmidt, M., and StLirwald. S., Kassel University Press, Kassel, Germany, 2008, pp. 605–617.

Liu, C.-T. and Huang, J.-S., "Fire Performance of Highly Flowable Reactive Powder Concrete," *Construction and Building Materials*, Vol. 23, No. 5, May 2009, pp. 2,072–2,079.

Lohaus, L. and Elsmeier, K., "Fatigue Behaviour of Plain and Fibre Reinforced Ultra-High Performance Concrete," *Proceedings of Hipermat 2012 3rd International Symposium on UHPC and Nanotechnology for High Performance Construction Materials*, Ed., Schmidt, M., Fehling, E., Glotzbach, C., Fröhlich, S., and Piotrowski, S., Kassel University Press, Kassel, Germany, 2012, pp. 631–637.

Lohaus, L. et al., "Experimental Analysis and Numerical Simulation of Ultra-High-Performance Concrete Tube Columns with a Steel Sheet Wrapping for Large Size Truss Structures," *Proceedings of Hipermat 2012 3rd International Symposium on UHPC and Nanotechnology for High Performance Construction Materials*, Ed., Schmidt, M., Fehling, E., Glotzbach, C., Fröhlich, S., and Piotrowski, S., Kassel University Press, Kassel, Germany, 2012, pp. 741–748.

López, J.A. et al., "Structural Design and Preliminary Calculations of a UHPFRC Truss Footbridge," *Proceedings of Hipermat 2012 3rd International Symposium on UHPC and*

Nanotechnology for High Performance Construction Materials, Ed., Schmidt, M., Fehling, E., Glotzbach, C., Fröhlich, S., and Piotrowski, S., Kassel University Press, Kassel, Germany, 2012, pp. 897–904.

Loukili, A., Richard, P., and Lamirault, J., "A Study on Delayed Deformations of an Ultra High Strength Cementitious Material," *Fourth CANMET/ACI/JCI Conference: Advances in Concrete Technology,* Publication No. SP-179, Ed., Malhotra, V.M., American Concrete Institute, Farmington Hills, MI, 1998, pp. 929–950.

Lubbers, A.R., "Bond Performance Between Ultra-High Performance Concrete and Prestressing Strands," Master's Thesis, College of Engineering and Technology, Ohio University, 2003.

Ma, J. and Schneider, H., "Properties of Ultra-High-Performance Concrete," in *LACER LEIPZIG Annual Civil Engineering Report*, Vol. 7, 2005, pp. 25–32.

Ma, J. et al., "Comparative Investigations on Ultra-High Performance Concrete With and Without Coarse Aggregates," *Proceedings of the International Symposium on Ultra High Performance Concrete*, Ed., Schmidt, M., Fehling, E., and GeisenhanslLike, C., Kassel University Press, Kassel, Germany, 2004, pp. 205–212.

Ma, J., Dietz, J., and Dehn, F., "Ultra High Performance Self Compacting Concrete," *Proceedings of the 3rd International RILEM Symposium on Self Compacting Concrete*, Publication pro033, Ed., Wallevik, O. and Nielsson, I., 2003, pp. 136–142.

Magnusson, J., Unosson, M., and Carlberg, A., "High Performance Concrete "HPC"—Field Experiments and Production," FOI, Swedish Defense Research Agency Technical Report, FOI-R-0256-SE, November 2001.

Maguire, M. et al., "Ultra-High-Performance Concrete in Standard Precast/Prestressed Concrete Products," *Proceedings of the PCI National Bridge Conference*, September 12–15, 2009, San Antonio, TX, Paper 33.

Magureanu, C. et al., "Mechanical Properties and Durability of Ultra-High-Performance Concrete," *ACI Materials Journal*, Vol. 109, No. 2, March–April 2012, pp. 177–183.

Malik, A.R. and Foster, S.J., "Behaviour of Reactive Powder Concrete Columns Without Steel Ties," *Journal of Advanced Concrete Technology*, Vol. 6, No. 2, 2008, pp. 377–386.

Marchand, P. et al., "Behaviour of an Orthotropic Bridge Deck with a UHPFRC Topping Layer," *Proceedings of Hipermat 2012 3rd International Symposium on UHPC and Nanotechnology for High Performance Construction Materials*, Ed., Schmidt, M., Fehling, E., Glotzbach, C., Fröhlich, S., and Piotrowski, S., Kassel University Press, Kassel, Germany, 2012, pp. 905–912.

Marchand, P. et al., "Modelling Flexural Tests on UHPFRC Thin-Walled Structures," *Proceedings of Hipermat 2012 3rd International Symposium on UHPC and Nanotechnology for High Performance Construction Materials*, Ed., Schmidt, M., Fehling, E., Glotzbach, C., Fröhlich, S., and Piotrowski, S., Kassel University Press, Kassel, Germany, 2012, pp. 757–764.

Marchand, P., "Béton Fibrés Ultra Performant (BFUP) Les Nouvelles Recommandations AFGC Chapitre 2, Calcul," (Updated AFGC Recommendations: Section 2 Structural Design), *Proceedings of the International Workshop on Ultra High Performance Fibre Reinforced Concrete—Designing and Building with UHPFRC: State of the Art Development*, 17–19 November 2009, Marseille, France, AFGC/*fib*, Paper 6.1.3.

Marchand, P., "Bétons Fibrés Ultra Performant (BFUP) et Développement Durable: Synthèse des Données Disponibles," (Available Data: An Overview), *Proceedings of the International Workshop on Ultra High Performance Fibre Reinforced Concrete—Designing and Building with UHPFRC: State of the Art Development*, 17–19 November 2009, Marseille, France, AFGC/*fib*, Paper 6.2.1.

Marijan, S., Dubravka, B., and Zeljana, S., "Acoustic Emission Response and Mechanical Characterization of Ultra High-Performance Concrete Types," *Seventh International Symposium on the Utilization of High-Strength/High-Performance Concrete*, Vol. II, Publication No. SP-228, Ed., Russell, H.G., American Concrete Institute, Farmington Hills, MI, 2005, pp. 1,207–1,220.

Markeset, G., "Ultra High Performance Concrete Is Ideal for Protective Structures," *Proceedings of the Third International Conference on High-Performance Concrete—Performance and Quality of Concrete Structures*, Recife, PE, Brazil, Publication No. SP-207, Ed., Malhotra, V.M. et al., American Concrete Institute, 2002, pp. 125–137.

Massicotte, B. and Boucher-Proulx, G., "Seismic Retrofitting of Bridge Piers With UHPFRC Jackets," *Proceedings of the International Workshop on Ultra High Performance Fibre Reinforced Concrete—Designing and Building with UHPFRC: State of the Art Development*, 17–19 November 2009, Marseille, France, AFGC/*fib*, Paper 5.1.1, 9 pp. Also *Designing and Building with UHPFRC: State of the Art and Development*, Ed., Resplendino, J. and Toutlemonde, F., pp. 531–540, Wiley-ISTE, London, 2010.

Massidda, L. et al., "High Pressure Steam Curing of Reactive-Powder Mortars," *Fifth CANMET/ACI Conference on Recent Advances in Concrete Technology, Fifth International Conference*, Publication No. SP-200, Ed., Malhotra, V.M., American Concrete Institute, Farmington Hills, MI, 2001, pp. 447–463.

Matsubara et al., "Application of a New Type of Ultra High Strength Fiber Reinforced Concrete to a Prestressed Concrete Bridge," *Proceedings of the Second International Symposium on Ultra High Performance Concrete*, Ed., Fehling, E., Schmidt, M., and Stürwald. S., Kassel University Press, Kassel, Germany, 2008, pp. 787–802.

Matte, V. et al., "Durability of Reactive Powder Concrete When Subjected to a Leaching Test," *Fifth CANMET/ACI Conference on Recent Advances in Concrete Technology, Fifth International Conference*, Publication No. SP-200, Ed., Malhotra, V.M., American Concrete Institute, Farmington Hills, MI, 2001, pp. 433–445.

Mazanec, O. and Schiep', P., "Mixing Time Optimization for UHPC," *Proceedings of the Second International Symposium on Ultra High Performance Concrete*, Ed., Fehling, E., Schmidt, M., and Stürwald. S., Kassel University Press, Kassel, Germany, 2008, pp. 401–408.

Mazanec, O., Lowke, D., and Schiep', P., "Mixing of High Performance Concrete: Effect of Concrete Composition and Mixing Intensity on Mixing Time," *Materials and Structure*, Vol. 43, No. 3, 2010, pp. 357–365.

Mazzacane, P., Ricciotti, R., and Teply, F., "La Passerelle des Anges," (The Angels Footbridge), *Proceedings of the International Workshop on Ultra High Performance Fibre Reinforced Concrete—Designing and Building with UHPFRC: State of the Art Development*, 17–19 November 2009, Marseille, France, AFGC/*fib*, Paper 3.1.1.

Meade, T.M. and Graybeal, B.A., "Flexural Response of Lightly Reinforced Ultra-High Performance Concrete Beams," *Proceedings of the Third International fib Congress and Exhibition Incorporating the PCI Annual Convention and National Bridge Conference*, Washington, DC, May 29–June 2, 2010, Compact Disc, Paper 101.

Meda, A. and Rosati, G., "Design and Construction of a Bridge in Very High Performance Fiber-Reinforced Concrete," *ASCE Journal of Bridge Engineering*, Vol. 8, No.5, 2003, pp. 281–287.

Millard, A. et al., "Dynamic Enhancement of Blast-Resistant Ultra High Performance Fiber-Reinforced Concrete Under Flexural and Shear Loading," *International Journal of Impact Engineering*, Vol. 37, No. 4, 2010, pp. 405–413.

Millon, O. et al., "Failure Mechanisms of UHPC Components Under Explosive Loading," *Proceedings of the Hipermat 2012 3rd International Symposium on UHPC and Nanotechnology for High Performance Construction Materials*, Ed., Schmidt, M., Fehling, E., Glotzbach, C., Fröhlich, S., and Piotrowski, S., Kassel University Press, Kassel, Germany, 2012, pp. 583–591.

Monai, B. and Schnabl, H., "Practice of UHPC in Austria," *Proceedings of the Second International Symposium on Ultra High Performance Concrete*, Ed., Fehling, E., Schmidt, M., and Stürwald. S., Kassel University Press, Kassel, Germany, 2008, pp. 839–846.

Moore, B. and Bierwagen, D., "Ultra High Performance Concrete Highway Bridge," *Proceedings of the 2006 Concrete Bridge Conference: HPC: Build Fast, Build to Last.* May 7–10, 2006, Reno, NV, Portland Cement Association, Skokie, IL, Compact Disc.

Moore, B.P. and Bierwagen, D., "Ultra High-Performance Concrete Highway Bridge," *Proceedings of the PCI National Bridge Conference*, October 23–25, 2006, Grapevine, TX, Compact Disc, Paper 54, 1 pp. Abstract only.

Moore, B., "Wapello County Looks to the Future," *ASPIRE*, Winter 2011, p. 46. Available at http://www.aspirebridge.org [Cited November 23, 2011].

Moore, B., "Little Cedar Creek Bridge—Big Innovation," *ASPIRE*, Spring 2012, p. 27. Available at http://www.aspirebridge.org [Cited April 20, 2012].

Morcous, G. and Tadros, M.K., *Applications of Ultra-High Performance Concrete in Bridge Girders*, Nebraska Department of Roads Report P310, January 2009.

Morcous, G., Maguire, M., and Tadros, M.K., "Shear Capacity of Ultra-High-Performance Concrete I-Girders with Orthogonal Welded Wire Reinforcement," *Proceedings of the Thomas T. C. Hsu Symposium: Shear and Torsion in Concrete Structures*, New Orleans, LA, Publication No. SP-265, American Concrete Institute, Farmington Hills, MI, 2009, Compact Disc, pp. 511–532.

Moreillon, L., Nseir, J., and Suter, R., "Shear and Flexural Strength of Thin UHPC Slabs," *Proceedings of Hipermat 2012 3rd International Symposium on UHPC and Nanotechnology for High Performance Construction Materials*, Ed., Schmidt, M., Fehling, E., Glotzbach, C., Fröhlich, S., and Piotrowski, S., Kassel University Press, Kassel, Germany, 2012, pp. 749–756.

Morris, A.D. and Garrett, G.G., "A Comparative Study of the Static and Fatigue Behaviour of Plain and Steel Fibre Reinforced Mortar in Compression and Direct Tension," *International Journal of Cement Composites and Lightweight Concrete*, Vol. 3, No. 2, 1981, pp. 73–91.

Muehlbauer, C. and Zilch, K., "Experimental Investigation of the Long-Term Behaviour of Glued UHPC Joints," *Proceedings of the Second International Symposium on Ultra High Performance Concrete*, Ed., Fehling, E., Schmidt, M., and Stürwald. S., Kassel University Press, Kassel, Germany, 2008, pp. 561–568.

Muehlbauer, C. and Zilch, K., "Glued Joints of Ultra High Performance Concrete Structures," *Proceedings of the Third International fib Congress and Exhibition Incorporating the PCI Annual Convention and National Bridge Conference*, Washington, DC, May 29–June 2, 2010, Compact Disc, Paper 41.

Müller, C. et al., "Durability of Ultra-High Performance Concrete (UHPC)," *Proceedings of the Third International fib Congress and Exhibition Incorporating the PCI Annual Convention and National Bridge Conference*, Washington, DC, May 29–June 2, 2010, Compact Disc, Paper 135.

Müller, C. et al., "Ultra-High Performance Spun Concrete Columns with High-Strength Reinforcement," *Proceedings of Hipermat 2012 3rd International Symposium on UHPC and Nanotechnology for High Performance Construction Materials*, Ed., Schmidt, M., Fehling, E., Glotzbach, C., Fröhlich, S., and Piotrowski, S., Kassel University Press, Kassel, Germany, 2012, pp. 733–740.

Müller, H.S. et al., "Time-Dependent Behaviour of Ultra High Performance Concrete (UHPC)," *Proceedings of the Third International fib Congress and Exhibition Incorporating the PCI Annual Convention and National Bridge Conference*, Washington, DC, May 29–June 2, 2010, Compact Disc, Paper 178.

Musha, H. et al., "Mikaneike Bridge—The First Continuous Girder Bridge Using Ultra-High Strength Fiber Reinforced Concrete," *Journal of Prestressed Concrete*, Japan, Vol. 49, No. 5, September–October, 2007, pp. 18–26.

Mutsuyoshi, H. et al., "Experimental Investigation of HFRP Composite Beams," *Fiber-Reinforced Polymer Reinforcement for Concrete Structures Tenth International Symposium CD-ROM,* Publication No. SP-275, Ed., Sen, R. et al., American Concrete Institute, Farmington Hills, MI, 2006, Compact Disc, pp. 1–26.

Naaman, A.E. and Wille, K., "The Path to Ultra-High Performance Fiber Reinforced Concrete (UHP-FRC): Five Decades of Progress," *Proceedings of Hipermat 2012 3rd International Symposium on UHPC and Nanotechnology for High Performance Construction Materials*, Ed., Schmidt, M., Fehling, E., Glotzbach, C., Fröhlich, S., and Piotrowski, S., Kassel University Press, Kassel, Germany, 2012, pp. 3–15.

Naaman, A.E., Reinhardt, H.W., and Fritz, C., "Reinforced Concrete Beams with SIFCON Matrix," *ACI Structural Journal*, Vol. 89, No. 1. January–February 1992, pp. 79–88.

Naaman, A., Likhitruangsilp, V., and Parra-Montesinos, G., "Punching Shear Response of High-Performance Fiber-Reinforced Cementitious Composite Slabs," *ACI Structural Journal*, Vol. 104, No. 2, March–April 2007, pp. 170–179.

New York State Department of Transportation, "557.21.16—Field Cast Joints Between Precast Concrete Units," July 2010.

Ngo, T., Mendis, P., and Krauthammer, T., "Behavior of Ultrahigh-Strength Prestressed Concrete Panels Subjected to Blast Loading," *Journal of Structural Engineering*, Vol. 133, No. 11, 2007, pp. 1,582–1,590.

Nielsen, C.V., "Tensile Postcrack Behavior of Steel Fiber Reinforced Ultra-High Strength Concrete," *International Workshop on High Performance Concrete,* Publication No. SP-159, Ed., Zia, P., American Concrete Institute, Farmington Hills, MI, 1996, pp. 231–246.

Nöldgen, M. et al., "Ultra High Performance Concrete Structures Under Aircraft Engineer Missile Impact," *Proceedings of Hipermat 2012 3rd International Symposium on UHPC and Nanotechnology for High Performance Construction Materials*, Ed., Schmidt, M., Fehling, E., Glotzbach, C., Fröhlich, S., and Piotrowski, S., Kassel University Press, Kassel, Germany, 2012, pp. 593–600.

O'Neil, E.F. and Dowd, W.M., "High-Performance Pipe Products Fabricated With Reactive Powder Concrete," *Proceedings of the Materials Engineering Conference*, 10-14 November 1996, Washington, DC, pp. 1,320–1,329.

O'Neil, E.F., Dauriac, C.E., and Gilliland, S.K., "Development of Reactive Powder Concrete (RPC) Products in the United States Construction Market," *High-Strength Concrete: An International Perspective*, Publication No. SP-167, Ed., Bickley, J.A., American Concrete Institute, Farmington Hills, MI, 1996, pp. 249–261.

Obata, H. et al., "Effect of Improved Interfacial Bond on Whitetopping Using Ultra High Strength Fiber Reinforced Concrete," *Proceedings of the Second International Symposium on Ultra High Performance Concrete*, Ed., Fehling, E., Schmidt, M., and Stürwald. S., Kassel University Press, Kassel, Germany, 2008, pp. 243–250.

Ocel, J. and Graybeal, B., "Fatigue Behavior of an Ultra-High Performance Concrete I-Girder," *Proceedings of the PCI National Bridge Conference*, October 22–24, 2007, Phoenix, AZ, Compact Disc, Paper 82.

Oesterlee, C., Sadouki, H., and Brühwiler, E., "Structural Analysis of a Composite Bridge Girder Combining UHPFRC and Reinforced Concrete," *Proceedings of the Second International Symposium on Ultra High Performance Concrete*, Ed., Fehling, E., Schmidt, M., and Stürwald. S., Kassel University Press, Kassel, Germany, 2008, pp. 647–654.

Okuma, H. et al., "The First Highway Bridge Applying Ultra High Strength Fiber Reinforced Concrete in Japan," *7th International Conference on Short and Medium Span Bridges*, Montreal, Canada, 2006.

Ono, T., "Application of Ultra-High-Strength Fiber Reinforced Concrete for Irrigation Channel Repair Works," *Proceedings of the International Workshop on Ultra High Performance Fibre Reinforced Concrete—Designing and Building with UHPFRC: State of the Art Development*, 17–19 November 2009, Marseille, France, AFGC/*fib*, Paper 5.1.2.

Orange, G., Dugat, J., and Acker, P., "DUCTAL: New Ultra-High-Performance Concretes. Damage, Resistance and Micromechanical Analysis," *Proceedings of the Fifth RILEM Symposium on Fibre-Reinforced Concretes (FRC) BEFIB' 2000*, Publication pro015, Lyon, France, Ed., Rossi, P. and Chanvillard, G., 2000, pp.781–790.

Orgass, M. and Klug, Y., "Fibre Reinforced Ultra-High Strength Concretes," *Proceedings of the International Symposium on Ultra High Performance Concrete*, Ed., Schmidt, M., Fehling, E., and GeisenhanslLike, C., Kassel University Press, Kassel, Germany, 2004, pp. 637–647.

Ozyildirim, C., "Evaluation of Ultra-High-Performance Fiber-Reinforced Concrete," Virginia Center for Transportation Innovation and Research, Report No. FHWA/VCTIR 12-R1, Federal Highway Administration, McLean, VA, 2011.

Ozyildirim, H.C. and Volgyi, J.F.J., "Virginia's Developments in the Use of Concrete in Bridges," *ASPIRE,* Winter 2008, pp. 50–52. Available at http://www.aspirebridge.org [Cited November 23, 2011].

Palecki, S. and Setzer, M.J., "Ultra-High-Performance Concrete Under Frost and De-Icing Salt Attack," *Proceedings of the Second International Symposium on Ultra High Performance Concrete*, Ed., Fehling, E., Schmidt, M., and StLirwald. S., Kassel University Press, Kassel, Germany, 2008, pp. 443–451.

Parant, E. et al., "Strain Rate Effect on Bending Behavior of New Ultra-High-Performance Cement-Based Composite," *ACI Materials Journal*, Vol. 104, No. 5, September-October 2007, pp. 458–463.

Park, H., "Model-Based Optimization of Ultra High Performance Concrete Highway Bridge Girders," Master's Thesis, Massachusetts Institute of Technology, 2003.

Park, H., Chuang, E., and Ulm, F.-J., "Model-Based Design of UHPC Highway Bridge Girders," *Proceedings of the 3rd International Symposium on High Performance Concrete/PCI National Bridge Conference*, October 19–22, 2003, Orlando, FL, 2003, Compact Disc, Paper 20, abstract only.

Park, S.H. et al., "Effect of Adding Micro Fibers on the Pullout Behavior of High Strength Steel Fibers in UHPC Matrix," *Proceedings of Hipermat 2012 3rd International Symposium on UHPC and Nanotechnology for High Performance Construction Materials*, Ed., Schmidt, M., Fehling, E., Glotzbach, C., Fröhlich, S., and Piotrowski, S., Kassel University Press, Kassel, Germany, 2012, pp. 541–548.

Park, S.Y. et al., "Structural Performance of Prestressed UHPC Ribbed Deck for Cable-Stayed Bridge," *Proceedings of Hipermat 2012 3rd International Symposium on UHPC and Nanotechnology for High Performance Construction Materials*, Ed., Schmidt, M., Fehling, E., Glotzbach, C., Fröhlich, S., and Piotrowski, S., Kassel University Press, Kassel, Germany, 2012, pp. 873–880.

Parra-Montesinos, G.J., Peterfreund, S., and Chao, S.-H., "Highly Damage-Tolerant Beam-Column Joints Through Use of High-Performance Fiber-Reinforced Cement Composites," *ACI Structural Journal*, Vol. 102, No. 3, May–June 2005, pp. 487–495.

Parsekian, G.A. et al., "Full-Scale Testing of a Fibre-Reinforced Concrete Footbridge," *Proceedings of the Institution of Civil Engineers: Bridge Engineering*, Vol. 162, No. 4, 2009, pp. 157–166.

Parsekian, G.A. et al., "Static and Fatigue Tests on Ductal® UHPFRC Footbridge Sections," *Proceedings of the Fifth ACI/CANMET/IBRACON International Conference on High-Performance Concrete Structures and Materials*, 18–20 June 2008, Manaus, Amazon State (AM), Brazil, Publication No. SP-253, Ed., Figueiredo, E.P. et al., American Concrete Institute, Farmington Hills, MI, 2008, pp. 273–290.

Perry, V. and Weiss, G., "Innovative Field Cast UHPC Joints for Precast Bridge Decks—Design, Prototype Testing and Projects," *Proceedings of the International Workshop on Ultra High Performance Fibre Reinforced Concrete—Designing and Building with UHPFRC: State of the Art Development*, 17–19 November 2009, Marseille, France, AFGC/*fib*, Paper 4.2.4.

Perry, V. and Seibert, P., "Field Cast UHPC Connections for Precast Bridge Elements and Systems," *Proceedings of Hipermat 2012 3rd International Symposium on UHPC and Nanotechnology for High Performance Construction Materials*, Ed., Schmidt, M.,

Fehling, E., Glotzbach, C., Fröhlich, S., and Piotrowski, S., Kassel University Press, Kassel, Germany, 2012, pp. 669–678.

Perry, V. and Zakariasen, D., "The First Use of UHPC Technology for an Innovative LRT Station Canopy Shawnessy, Calgery, Alberta," *Seventh International Symposium on the Utilization of High-Strength/High-Performance Concrete*, Vol. II, Publication No. SP-228, Ed., Russell, H.G., American Concrete Institute, Farmington Hills, MI, 2005, pp. 1,119–1,131.

Perry, V.H. and Royce, M., "Innovative Field-Cast UHPC Joints for Precast Decks (Side-by-Side Deck Bulb-Tees),Village of Lyons, New York: Design, Prototyping, Testing and Construction," *Proceedings of the Third International fib Congress and Exhibition Incorporating the PCI Annual Convention and National Bridge Conference*, Washington, DC, May 29–June 2, 2010, Compact Disc, Paper 401, 13 pp.

Perry, V.H. and Seibert, P.J., "New Applications of Field Cast UHPC Connections for Precast Bridges," *Proceedings of the PCI National Bridge Conference*, October 22–26, 2011, Salt Lake City, UT, Compact Disc, Paper 101.

Perry, V.H. and Seibert, P.J., "The Use of UHPFRC (Ductal®) for Bridges in North America: The Technology, Applications and Challenges Facing Commercialization," *Proceedings of the Second International Symposium on Ultra High Performance Concrete*, Ed., Fehling, E., Schmidt, M., and Stürwald. S., Kassel University Press, Kassel, Germany, 2008, pp. 815–822. Also available at http://www.ductal-lafarge.com/SLib/17-UHPC [Cited November 23, 2011].

Perry, V.H. and Zakariasen, D., "Overview of UHPC Technology, Materials, Properties, Markets and Manufacturing," *Proceedings of the 3rd International Symposium on High Performance Concrete/PCI National Bridge Conference*, October 19–22, 2003, Orlando, FL, Compact Disc, Paper 60.

Perry, V.H., "Q&A: What Is Reactive Powder Concrete," *HPC Bridge Views*, Issue 16, July/August 2001. Available at http://www.hpcbridgeviews.org [Cited November 23, 2011].

Perry, V., Moore, B., and Bierwagen, D., "Revolutionary Concrete Solutions—Understanding Ultra-High-Performance, Fiber-Reinforced Concrete," *Construction Specifier,* Vol. 59, No. 10, 2006, pp. 40–52.

Perry, V., Scalzo, P., and Weiss, G., "Innovative Field Cast UHPC Joints for Precast Deck Panel Bridge Superstructures—CN Overhead Bridge at Rainy Lake, Ontario," *Proceedings of the PCI National Bridge Conference*, October 22–24, 2007, Phoenix, AZ, Compact Disc, Paper 3.

Peters, H., Bäziger, H., and Poltera, M, "Durable Adhesive Bonding with Epoxy Resins in Civil Engineering Construction," *Proceedings of the Second International Symposium on Ultra High Performance Concrete*, Ed., Fehling, E., Schmidt, M., and Stürwald. S., Kassel University Press, Kassel, Germany, 2008, pp. 267–274.

Petitjean, J, and Resplendino, J., "French Recommendations for Ultra-High Performance Fiber-Reinforced Concretes," *Proceedings of the 6th International Symposium on High Strength/High Performance Concrete*, Leipzig, Germany, June 2002, Ed., König, G., Dehn, F., and Faust, T., Vol. 1, pp. 485–498.

Pfeifer, C. et al., "Durability of Ultra-High-Performance Concrete," *10th ACI International Conference on Recent Advances in Concrete Technology and Sustainability Issues,*

Publication No. SP-261, Ed., Gupta, P., Holland, T.C., and Malhotra, V.M., American Concrete Institute, Farmington Hills, MI, 2006, pp. 1–16.

Piérard, J., Dooms, B., and Cauberg, N., "Evaluation of Durability Parameters of UHPC Using Accelerated Lab Tests," *Proceedings of Hipermat 2012 3rd International Symposium on UHPC and Nanotechnology for High Performance Construction Materials*, Ed., Schmidt, M., Fehling, E., Glotzbach, C., Fröhlich, S., and Piotrowski, S., Kassel University Press, Kassel, Germany, 2012, pp. 371–376.

Pimienta, P. et al., "Behaviour of UHPFRC at High Temperatures," *Proceedings of the International Workshop on Ultra High Performance Fibre Reinforced Concrete—Designing and Building with UHPFRC: State of the Art Development*, 17–19 November 2009, Marseille, France, AFGC/*fib*, Paper 5.3.2.

Pimienta, P. et al., "Literature Review on the Behavior of UHPFRC at High Temperature," *Proceedings of Hipermat 2012 3rd International Symposium on UHPC and Nanotechnology for High Performance Construction Materials*, Ed., Schmidt, M., Fehling, E., Glotzbach, C., Fröhlich, S., and Piotrowski, S., Kassel University Press, Kassel, Germany, 2012, pp. 549–556.

Piotrowski, S. and Schmidt, M., "Life Cycle Cost Analysis of a UHPC-Bridge on Example of Two Bridge Refurbishment Designs," *Proceedings of Hipermat 2012 3rd International Symposium on UHPC and Nanotechnology for High Performance Construction Materials*, Ed., Schmidt, M., Fehling, E., Glotzbach, C., Fröhlich, S., and Piotrowski, S., Kassel University Press, Kassel, Germany, 2012, pp. 957–964.

Plank, J., Schröfl, C., and Gruber, M., "Use of a Supplemental Agent to Improve Flowability of Ultra-High-Performance Concrete," *Ninth ACI International Conference on Superplasticizers and Other Chemical Admixtures*, Publication No. SP-262, Ed., Gupta, P., Holland, T.C., and Malhotra, V.M., American Concrete Institute, Farmington Hills, MI, 2009, pp. 1–16.

Qian, S. and Li, V.C., "Simplified Inverse Method for Determining the Tensile Properties of SHCCs," *Journal of Advanced Concrete Technology*, Vol. 6, No. 2, 2008, pp. 353–363.

Racky, P., "Cost-Effectiveness and Sustainability of UHPC," *Proceedings of the International Symposium on Ultra High Performance Concrete*, Ed., Schmidt, M., Fehling, E., and GeisenhanslLike, C., Kassel University Press, Kassel, Germany, 2004, pp. 797–805.

Rafiee, A. and Schmidt, M., "Computer Modeling and Investigation on the Chloride Induced Steel Corrosion in Cracked UHPC," *Proceedings of Hipermat 2012 3rd International Symposium on UHPC and Nanotechnology for High Performance Construction Materials*, Ed., Schmidt, M., Fehling, E., Glotzbach, C., Fröhlich, S., and Piotrowski, S., Kassel University Press, Kassel, Germany, 2012, pp. 357–364.

Rahman, S., Molyneaux, T., and Patnaikuni, I., "Ultra High Performance Concrete: Recent Applications and Research," *Australian Journal of Civil Engineering*, Vol. 2, No. 1, 2005, pp. 13–20.

Rajlic, B. et al., "The Eagle River Bridge Superstructure Replacement," *Proceedings of the 8th International Conference on Short and Medium Span Bridges,* The Canadian Society for Civil Engineering, Niagara Falls, ON, 2010.

Randl, N. et al., "Study on the Application of UHPC for Precast Tunnel Segments," *Proceedings of Hipermat 20123rd International Symposium on UHPC and*

Nanotechnology for High Performance Construction Materials, Ed., Schmidt, M., Fehling, E., Glotzbach, C., Fröhlich, S., and Piotrowski, S., Kassel University Press, Kassel, Germany, 2012, pp. 981–988.

Rebentrost, M. and Wight, G., "Experience and Applications of Ultra-High Performance Concrete in Asia," *Proceedings of the Second International Symposium on Ultra High Performance Concrete*, Ed., Fehling, E., Schmidt, M., and StLirwald. S., Kassel University Press, Kassel, Germany, 2008, pp. 19–30.

Rebentrost, M. and Wight, G., "Investigation of UHPFRC Slabs Under Blast Loads," *Proceedings of the International Workshop on Ultra High Performance Fibre Reinforced Concrete—Designing and Building with UHPFRC: State of the Art Development*, 17–19 November 2009, Marseille, France, AFGC/*fib*, Paper 4.1.6.

Rebentrost, M. and Wight, G., "UHPC Perspective From a Specialist Construction Company," *Proceedings of the International Workshop on Ultra High Performance Fibre Reinforced Concrete—Designing and Building with UHPFRC: State of the Art Development*, 17–19 November 2009, Marseille, France, AFGC/*fib*, Paper 3.1.5.

Rebentrost, M., "Australian Experience with Ductal®: An Ultra-High Performance Concrete," *Proceedings of the Second International fib Congress*, June 5–8, 2006, Naples, Italy. Available on Compact Disc from http://www.fib-international.org/proceedings-of-previous-conferences [Cited November 23, 2011].

Rebentrost, M., "Design and Construction of the First Ductal Bridge in New Zealand," *Proceedings of the New Zealand Concrete Industry Conference*, September 2005, Auckland, New Zealand.

Rebentrost, M. and Wight, G., "Behaviour and Resistance of Ultra High Performance Concrete to Blast Effects," *Proceedings of the 2nd International Symposium on Ultra-High Performance Concrete*, Kassel, Germany, 2008, pp. 735–742.

Reda Taha, M.M. and Shrive, N.G., "New Concrete Anchors for Carbon Fiber-Reinforced Polymer Post-Tensioning Tendons—Part 1: State-of-the-Art Review/Design," *ACI Structural Journal*, Vol. 100, No. 1, January–February 2003, pp. 86–95.

Reda Taha, M.M. and Shrive, N.G., "New Concrete Anchors for Carbon Fiber-Reinforced Polymer Post-Tensioning Tendons—Part 2: Development/Experimental Investigation," *ACI Structural Journal*, Vol. 100, No. 1, January–February 2003, pp. 96–104.

Reda Taha, M.M. and Shrive, N.G., "UHPC Anchors for Post-Tensioning. Ultra-High-Performance Concrete Anchors for CFRP Tendons Can Be a Step Towards Metal-Free Structures," *Concrete International*, Vol. 25, No. 8, August 2003, pp. 35–40.

Reeves, E.E., "Structural Reliability of Ultra-High Performance Concrete in Flexure," Master's Thesis, College of Engineering and Technology, Ohio University, Athens, OH, 2004.

Reichel, M. et al., "Road Bridge WILD—UHPFRC for a Segmental Arch Structure," *Proceedings of the International Workshop on Ultra High Performance Fibre Reinforced Concrete—Designing and Building with UHPFRC: State of the Art Development*, 17–19 November 2009, Marseille, France, AFGC/*fib*, Paper 3.2.1.

Reichel, M., Altersberger, G., and Sparowitz, L., "UHPFRC-Prototype for a Flexible Modular Temporary High Speed Railway Bridge," *Proceedings of the International Workshop on Ultra High Performance Fibre Reinforced Concrete—Designing and Building with*

UHPFRC: State of the Art Development, 17–19 November 2009, Marseille, France, AFGC/*fib*, Paper 3.2.5.

Reichel, M., Sparowitz, L., and Freytag, B., "UHPC-Segmental Bridges—Material-Based Design Principles and Adapted Construction Methods," *Proceedings of the Second International Symposium on Ultra High Performance Concrete*, Ed., Fehling, E., Schmidt, M., and Stürwald. S., Kassel University Press, Kassel, Germany, 2008, pp. 779–786.

Reineck, K.-H. and Frettlöhr, B., "Tests on Scale Effect of UHPFRC Under Bending and Axial Forces," *Proceedings of the Third International fib Congress and Exhibition Incorporating the PCI Annual Convention and National Bridge Conference*, Washington, DC, May 29–June 2, 2010, Compact Disc, Paper 54.

Resplendino, J. and Petitjean, J., "Ultra-High-Performance Concrete: First Recommendations and Examples of Application," *Proceedings of the 3rd International Symposium on High Performance Concrete/PCI National Bridge Conference*, October 19–22, 2003, Orlando, FL, Compact Disc, Paper 77.

Resplendino, J. and Toutlemonde, F., "Using UHPFRC for Design and Building: Gained Knowledge and Perspectives for Research Needs, Design Trends and Improvement of Construction Techniques for Sustainable UHPFRC Construction," *Proceedings of the International Workshop on Ultra High Performance Fibre Reinforced Concrete—Designing and Building with UHPFRC: State of the Art Development*, 17–19 November 2009, Marseille, France, AFGC/*fib*.

Resplendino, J. et al., "Construction of an Overpass on the A51 Motorway, Made of a Prestressed Box Beam Built with UHPFRC in the French Technology of Concrete," *Proceedings of the Second fib Congress*, June 5–8, 2006, Naples, Italy. Available on Compact Disc from http://www.fibinternational.org/proceedings-of-previous-conferences [Cited November 23, 2011].

Resplendino, J., "Bétons Fibrés Ultra Performant (BFUP)—Les Nouvelles Recommandations AFGC," (Updated AFGC Recommendations), *Proceedings of the International Workshop on Ultra High Performance Fibre Reinforced Concrete—Designing and Building with UHPFRC: State of the Art Development*, 17–19 November 2009, Marseille, France, AFGC/*fib*, Paper 6.1.1.

Resplendino, J., "First Recommendations for Ultra-High-Performance Concretes and Examples of Application," *Proceedings of the International Symposium on Ultra High Performance Concrete*, Ed., Schmidt, M., Fehling, E., and GeisenhanslLike, C., Kassel University Press, Kassel, Germany, 2004, pp. 79–90.

Resplendino, J., "Introduction: Qu'est ce qu'un BFUP," (Introduction: What Is UHPFRC?) *Proceedings of the International Workshop on Ultra High Performance Fibre Reinforced Concrete—Designing and Building with UHPFRC: State of the Art Development*, 17–19 November 2009, Marseille, France, AFGC/*fib*, Paper 1.1.

Resplendino, J., "State of the Art of Design and Construction of UHPFRC Structures in France," *Proceedings of Hipermat 2012 3rd International Symposium on UHPC and Nanotechnology for High Performance Construction Materials*, Ed., Schmidt, M., Fehling, E., Glotzbach, C., Fröhlich, S., and Piotrowski, S., Kassel University Press, Kassel, Germany, 2012, pp. 27–41.

Resplendino, J., "Ultra-High Performance Concretes—Recent Realizations and Research Programs on UHPFRC Bridges in France," *Proceedings of the Second International*

Symposium on Ultra High Performance Concrete, Ed., Fehling, E., Schmidt, M., and StLirwald. S., Kassel University Press, Kassel, Germany, 2008, pp. 31–43.

Ricciotti, R., "The Footbridge of Peace, Seoul, South Korea," *Concrete*, Vol. 36, No. 10, November/December 2002, pp. 11–13.

Richard, P. and Cheyrezy, M.H., "Reactive Powder Concretes with High Ductility and 200–800 MPa Compressive Strength," *Concrete Technology Past, Present, and Future, Proceedings of the V. Mohan Malhotra Symposium,* Publication No. SP-144, Ed., Mehta, P.K., American Concrete Institute, Farmington Hills, MI, 1994, pp. 507–518.

Richard, P. and Cheyrezy, M., "Composition of Reactive Powder Concretes," *Cement and Concrete Research*, Vol. 25, No. 7, 1995, pp. 1,501–1,511.

Richard, P., "Reactive Powder Concrete: A New-Ultra High Strength Cementitious Material," *Proceedings of the Fourth International Symposium on the Utilization of High-Strength/HighPerformance Concrete*, 29–31 May 1996, Paris, France, Ed., de Larrard, F. and Lacroix, R., Vol. 3, pp. 1,343–1,357.

RILEM TC 162-TDF, "Test and Design Methods for Steel Fibre Reinforced Concrete, Recommendations," *Materials and Structures*, Vol. 35, 2002, pp. 579–582.

Rivillon, P., Portelatine, J., and Nicolas, F., "Expérimentation et Modélisation de Poteaux Droits et de Forme Y en BFUHP Précontraints sous l'Action d'un Chargement Vertical Excentré: Projet du Musée des Civilisations de l'Europe et de la Méditerranée (MuCEM)," (Experimentation and Analysis of Straight and Y-Shaped Prestressed Columns Made of UHPFRC Under Vertical Off-Centered Loading), *Proceedings of the International Workshop on Ultra High Performance Fibre Reinforced Concrete—Designing and Building with UHPFRC: State of the Art Development*, 17–19 November 2009, Marseille, France, AFGC/*fib*, Paper 7.3.

Rossi, P. and Parant, E., "Durability of an Ultra-High Fibre Reinforced Cement Composite Under a Service State Loading and in an Aggressive Environment," *Proceedings of the International Conference on Cement Combinations for Durable Concrete*, 5–7 July 2005, Dundee, Scotland, Ed., Dhir, R.K., Harrison, T.A., and Newlands, M.D., pp. 99–106.

Rossi, P. et al., "Bending and Compressive Behaviors of a New Cement Composite," *Cement and Concrete Research*, Vol. 35, No. 1, 2005, pp. 27–33.

Rossi, P. et al., "Ultra-High-Strength Steel Fibre Reinforced Concretes: Mix Design and Mechanical Characterization," *Fiber Reinforced Concrete: Modern Developments*, Ed., Banthia, N. and Mindess, S., University of British Columbia, Vancouver, Canada, 1995, pp. 181–186.

Rossi, P., "High Performance Multimodal Fiber Reinforced Cement Composites (HPMFRCC): The LCPC Experience," *ACI Materials Journal*, Vol. 94, No. 6, November–December 1997, pp. 478–483.

Rossi, P., "Ultra High-Performance Concrete," *Concrete International*, Vol. 30, No. 2, February 2008, pp. 31–34.

Rossi, P., "Ultra-High Performance Fibre Reinforced Concretes (UHPFRC): An Overview," *Proceedings of the Fifth RILEM Symposium on Fibre-Reinforced Concretes (FRC) BEFIB' 2000*, Publication pro015, Lyon, France, Ed., Rossi, P. and Chanvillard, G., 2000, pp. 87–100.

Rossi, P., "Ultra-High-Performance Fiber-Reinforced Concretes. A French Perspective on Approaches Used to Produce High-Strength, Ductile Fiber-Reinforced Concrete," *Concrete International*, Vol. 23, No. 12, December 2001, p. 46–52.

Roth, J. et al., "Laboratory Investigation of the Characterization of Cor-Tuf Flexural and Splitting Tensile Properties," Technical Report ERDC/GSL TR-10-46, U.S. Army Corps of Engineers, Engineer Research and Development Center, Washington, DC, 2010.

Roth, M. et al., "Ultra-High Strength, Glass Fiber-Reinforced Concrete: Mechanical Behavior and Numerical Modeling," *ACI Materials Journal*, Vol.107, No. 2, March–April 2010, pp. 185–194.

Rougeau, P. and Borys, B., "Ultra High Performance Concrete with Ultrafine Particles Other Than Silica Fume," *Proceedings of the International Symposium on Ultra High Performance Concrete*, Ed., Schmidt, M., Fehling, E., and Geisenhanslüke, C., Kassel University Press, Kassel, Germany, 2004, pp. 213–225.

Roux, N., Andrade, C., and Sanjuan, M.A., "Experimental Study of Durability of Reactive Powder Concretes," *Journal of Materials in Civil Engineering*, 1996, Vol. 8, No. 1, pp. 1–6.

Roy, N., Proulx, J., and Paultre, P., "Dynamic Testing of the Sherbrooke Pedestrian Bridge," *Proceedings of the 2000 Annual Conference Abstracts*—Canadian Society for Civil Engineering, 7–10 June 2000, London, Ontario, p. 211.

Royce, M.C., "Concrete Bridges in New York State," *ASPIRE*, Fall 2011, pp. 46–48. Available at http://www.aspirebridge.org. [Cited November 23, 2011].

Ruiz, E.D. et al, "Transfer and Development Lengths of Prestressed Beams Cast with Ultra-High Performance Concrete," *Proceedings of the PCI National Bridge Conference*, October 4–7, 2008, Orlando, FL, Compact Disc, Paper 22.

Ruiz, E.D. et al., "Preliminary Investigation into the Use of UHPC in Prestressed Members," *Proceedings of the PCI National Bridge Conference*, October 22–24, 2007, Phoenix, AZ, Compact Disc, Paper 53.

Sachstandbericht, "Ultrahochfester Beton," Berlin, Beuth Verlag In: Schriftenreihe des Deutschen Ausschuss fur Stahlbeton, Nr. 561, Berlin, Germany, 2008.

Šajna, A., Denarié, E., and Bras, V., "Assessment of a UHPFRC Based Bridge Rehabilitation in Slovenia, Two Years After Application," *Proceedings of Hipermat 2012 3rd International Symposium on UHPC and Nanotechnology for High Performance Construction Materials*, Ed., Schmidt, M., Fehling, E., Glotzbach, C., Fröhlich, S., and Piotrowski, S., Kassel University Press, Kassel, Germany, 2012, pp. 937–944.

Saleem, M.A. et al., "Ultra-High Performance Concrete Light-Weight Deck for Moveable Bridges," *Proceedings of the Third International fib Congress and Exhibition Incorporating the PCI Annual Convention and National Bridge Conference*, Washington, DC, May 29–June 2, 2010, Compact Disc, Paper 361.

Saleem, M.A. et al., "Ultra-High-Performance Concrete Bridge Deck Reinforced With High-Strength Steel," *ACI Structural Journal*, Vol. 108, No. 5, September–October 2011, pp. 601–609.

SAMARIS, "Sustainable and Advanced Materials for Road Infra Structures," Brochure SAM_DE03, 2003, 8 pp. PDF available at http://samaris.zag.si/ [Cited November 23, 2011].

Sameer, S., "Behavioral Study of Ultra High Performance Concrete Girders," Master's Thesis, Department of Civil and Environmental Engineering, University of Maryland, College Park, 2004.

Sameer, S., Fu, C.C., (or Chung, C.F.) and Graybeal, B., "Behavioral Study of Ultra High Performance Concrete Girders," *Proceedings of the 3rd International Conference Construction Materials: Performance, Innovations and Structural Implications*, August 22–24, 2005, Vancouver, British Columbia.

Schachinger, I., Hilbig, H., and Stengal, T., "Effect of Curing Temperatures at an Early Age on the Long-Term Strength Development of UHPC," *Proceedings of the Second International Symposium on Ultra High Performance Concrete*, Ed., Fehling, E., Schmidt, M., and StLirwald. S., Kassel University Press, Kassel, Germany, 2008, pp. 205–212.

Schäfers, M. and Seim, W., "Development of Adhesive-Bound UHPC-Timber Composites," *Proceedings of the Second International Symposium on Ultra High Performance Concrete*, Ed., Fehling, E., Schmidt, M., and StLirwald. S., Kassel University Press, Kassel, Germany, 2008, pp. 259–266.

Scheffler, B. and Schmidt, M., "Application of UHPC for Multifunctional Road Pavements," *Proceedings of Hipermat 2012 3rd International Symposium on UHPC and Nanotechnology for High Performance Construction Materials*, Ed., Schmidt, M., Fehling, E., Glotzbach, C., Fröhlich, S., and Piotrowski, S., Kassel University Press, Kassel, Germany, 2012, pp. 913–920.

Scheydt, J.C. and MLiller H.S., "Microstructure of Ultra High Performance Concrete (UHPC) and Its Impact on Durability," *Proceedings of Hipermat 2012 3rd International Symposium on UHPC and Nanotechnology for High Performance Construction Materials*, Ed., Schmidt, M., Fehling, E., Glotzbach, C., Fröhlich, S., and Piotrowski, S., Kassel University Press, Kassel, Germany, 2012, pp. 349–356.

Schmidt, C. and Schmidt, M., "'Whitetopping' of Asphalt and Concrete Pavements with Thin Layers of Ultra-High-Performance Concrete—Construction and Economic Efficiency," *Proceedings of Hipermat 2012 3rd International Symposium on UHPC and Nanotechnology for High Performance Construction Materials*, Ed., Schmidt, M., Fehling, E., Glotzbach, C., Fröhlich, S., and Piotrowski, S., Kassel University Press, Kassel, Germany, 2012, pp. 921–927.

Schmidt, C. et al., "Strengthening and Rehabilitation of Pavements Applying Thin Layers of Reinforced Ultra-High-Performance Concrete (UHPC-White Topping)," *Proceedings of the Second International Symposium on Ultra High Performance Concrete*, Ed., Fehling, E., Schmidt, M., and Stürwald. S., Kassel University Press, Kassel, Germany, 2008, pp. 655–662.

Schmidt, D. and Fehling, E., "Ultra-High-Performance Concrete: Research, Development and Application in Europe," *Seventh International Symposium on the Utilization of High-Strength/High-Performance Concrete*, Vol. I, Publication No. SP-228, Ed., Russell, H.G., American Concrete Institute, Farmington Hills, MI, 2005, pp. 51–78.

Schmidt, M. and Fröhlich, S., "Testing of UHPC," *Proceedings of the Third International fib Congress and Exhibition Incorporating the PCI Annual Convention and National Bridge Conference*, Washington, DC, May 29–June 2, 2010, Compact Disc, Paper 237.

Schmidt, M. and Jerebic D., "UHPC: Basis for Sustainable Structures—The Gaertnerplatz Bridge in Kassel," *Proceedings of the Second International Symposium on Ultra High*

Performance Concrete, Ed., Fehling, E., Schmidt, M., and Stürwald. S., Kassel University Press, Kassel, Germany, 2008, pp. 619–625.

Schmidt, M. et al., "Durability of Ultra High Performance Concrete," *Proceedings of the 6th International Symposium on High Strength/High Performance Concrete*, Leipzig, Germany, June 2002, Ed., König, G., Dehn, F., and Faust, T., Vol. 2, pp. 1,367–1,376.

Schmidt, M. et al., "Mix Design and Properties of Ultra-High Performance Fibre Reinforced Concrete for the Construction of a Composite Concrete UHPFRC-Concrete Bridge," *Proceedings of IABSE Symposium Improving Infrastructure Worldwide—Bringing People Closer*, IABSE, Weimar, Germany, Vol. 93, 2007, pp. 466–476. Also *Ultra High Performance Concrete (UHPC)— 10 Years of Research and Development at the University of Kassel*, Ed., Schmidt, M. and Fehling, E., Kassel University Press, Kassel, Germany, 2007, pp. 44–54.

Schmidt, M., "Sustainable Building with Ultra-High-Performance Concrete (UHPC)—Coordinated Research Program in Germany," *Proceedings of the Third International fib Congress and Exhibition Incorporating the PCI Annual Convention and National Bridge Conference*, Washington, DC, May 29–June 2, 2010, Compact Disc, Paper 682.

Schmidt, M., "Sustainable Building with Ultra-High-Performance Concrete (UHPC)—Coordinated Research Program in Germany," *Proceedings of Hipermat 2012 3rd International Symposium on UHPC and Nanotechnology for High Performance Construction Materials*, Ed., Schmidt, M., Fehling, E., Glotzbach, C., Fröhlich, S., and Piotrowski, S., Kassel University Press, Kassel, Germany, 2012, pp. 17–25.

Schmidt, M., "Ultra-High Performance Concrete in Germany and the World—Current State of Research, Technical Rules and Standards and Practical Application," *Betonwerk und FertigteilTechnik/Concrete Plant and Precast Technology*, 2009, Vol. 75, No. 2, pp. 14–16.

Schmidt, M., et al., "Ultra-High Performance Concrete: Perspective for the Precast Concrete Industry," *Concrete Precasting Plant and Technology*, Vol. 69, No. 3, 2003, pp. 16–29.

Schneider, U., Diederichs, U., and Horvath, J., "Behaviour of Ultra High Performance Concrete (UHPC) Under Fire Exposure," *Beton- und Stahlbetonbau*, Vol. 98, No. 7, 2003, pp. 408–417.

Schnellenbach-Held, M. and Prager, M., "Numerical Study on the Shear Behavior of Micro-Reinforced UHPC Beams," *Proceedings of Hipermat 2012 3rd International Symposium on UHPC and Nanotechnology for High Performance Construction Materials*, Ed., Schmidt, M., Fehling, E., Glotzbach, C., Fröhlich, S., and Piotrowski, S., Kassel University Press, Kassel, Germany, 2012, pp. 469–476.

Sedran, T., Durand, C., and de Larrard, F., "An Example of UHPFRC Recycling," *Proceedings of the International Workshop on Ultra High Performance Fibre Reinforced Concrete—Designing and Building with UHPFRC: State of the Art Development*, 17–19 November 2009, Marseille, France, AFGC/*fib*, Paper 6.2.2.

Semioli, W.J., "The New Concrete Technology," *Concrete International*, Vol. 23, No. 11, November 2001, pp. 75–79.

Shaheen, E. and Shrive, N.G., "Reactive Powder Concrete Anchorage for Post-Tensioning With Carbon Fiber-Reinforced Polymer Tendons," *ACI Materials Journal*, Vol. 103, No. 6, November–December 2006, pp. 436–443.

Shann, S.V. et al., "Application of Ultra-High Performance Concrete (UHPC) as a Thin-Topped Overlay for Concrete Bridge Decks," *Proceedings of Hipermat 2012 3rd International Symposium on UHPC and Nanotechnology for High Performance Construction Materials*, Ed., Schmidt, M., Fehling, E., Glotzbach, C., Fröhlich, S., and Piotrowski, S., Kassel University Press, Kassel, Germany, 2012, pp. 929–936.

Shutt, C.A., "UHPC Joint Provides New Solutions," *ASPIRE*, Fall 2009, pp. 28–30. Available at http://www.aspirebridge.org [Cited November 23, 2011].

Simon A. et al., "Realization of Two Road Bridges With Ultra-High-Performance Fibre-Reinforced Concrete," *Proceedings of the 6th International Symposium on High Strength/High Performance Concrete*, Leipzig, Germany, June 2002, Ed., König, G., Dehn, F., and Faust, T., Vol. 1, pp. 753–768.

Simon, A., "Les Nouvelles Recommandations AFGC sur les BFUP CHAPITRE I—Comportement et Caractéristiques Mécaniques des BFUP," (Updated AFGC Recommendations: Section 1 Materials), *Proceedings of the International Workshop on Ultra High Performance Fibre Reinforced Concrete—Designing and Building with UHPFRC: State of the Art Development*, 17–19 November 2009, Marseille, France, AFGC/*fib*, Paper 6.1.2.

Skazlic, M.B. and Mikulic, D., "Optimized Composition of UHPC for Tunnel Lining," *Proceedings from 18th BIBM International Congress, Meet the Future of Precast Concrete,* 11–14 May, 2005. Amsterdam, Holland, BFBN, pp. 89–90.

Skazlić, M., Bjegović, D., and Tvrtkovic, D., "Reactive Powder Concrete for Better Earthquake Resistance," *Proceedings on the fib Symposium on Concrete Structures in Seismic Regions*, May 6-9, 2003, Athens, Greece 8 pp. Available at http://info.grad.hr/gf/index.asp?pid=1342&o=17 [Cited November 23, 2011].

Skazlić, M., Serdar, M., and Bjegović D., "Influence of Test Specimens Geometry on Compressive Strength of Ultra High Performance Concrete," *Proceedings of the Second International Symposium on Ultra High Performance Concrete*, Ed., Fehling, E., Schmidt, M., and Stürwald. S., Kassel University Press, Kassel, Germany, 2008, pp. 295–301.

Soh, M., "Model-Based Design of an Ultra High Performance Concrete Prototype Highway Bridge Girder," Master's Thesis, Massachusetts Institute of Technology, 2003.

Soliman, A. and Nehdi, M., "Early-Age Shrinkage of Ultra-High-Performance Concrete Under Drying/Wetting Cycles and Submerged Conditions," *ACI Materials Journal*, Vol. 109, No. 2, March–April 2012, pp. 131–139.

Sorelli, L. et al., "Risk Analysis of Early-Age Cracking in UHPC Structures," *Proceedings of the Second International Symposium on Ultra High Performance Concrete*, Ed., Fehling, E., Schmidt, M., and Stürwald. S., Kassel University Press, Kassel, Germany, 2008, pp. 331–338.

Sparowitz, L. and Reichel, M. M., "Fiber-Reinforced Ultra-High-Performance Concrete in Practice—Precast Bridge in Völkermarkt—Concept and Design," *Betonwerk und FertigteilTechnik/Concrete Plant and Precast Technology*, Vol. 75, No. 2, 2009, p. 44–45.

Sritharan, S. et al., "Current Research on Ultra High Performance Concrete (UHPC) for Bridge Applications in Iowa," *Proceedings of Hipermat 2012 3rd International Symposium on UHPC and Nanotechnology for High Performance Construction Materials*, Ed., Schmidt, M., Fehling, E., Glotzbach, C., Fröhlich, S., and Piotrowski, S., Kassel University Press, Kassel, Germany, 2012, pp. 857–864.

Sritharan, S., Bristow, B.J., and Perry, V.H., "Characterizing an Ultra-High Performance Material for Bridge Applications under Extreme Loads," *Proceedings of the 3rd International Symposium on High Performance Concrete/PCI National Bridge Conference*, October 19–22, 2003, Orlando, FL, Compact Disc, Paper 99.

Sritharan, S., Vande Voort, T., and Suleiman, M., "Effective Use of UHPC for Deep Foundation Piles," *Proceedings of the International Workshop on Ultra High Performance Fibre Reinforced Concrete—Designing and Building with UHPFRC: State of the Art Development*, 17–19 November 2009, Marseille, France, AFGC/*fib*, Paper 3.2.6.

Staquet, S. and Espion, B., "Early-Age Autogenous Shrinkage of UHPC Incorporating Very Fine Fly Ash or Metakaolin in Replacement of Silica Fume," *Proceedings of the International Symposium on Ultra High Performance Concrete*, Ed., Schmidt, M., Fehling, E., and GeisenhanslLike, C., Kassel University Press, Kassel, Germany, 2004, pp. 587–599.

Steinberg, E. and Lubbers, A., "Bond of Prestressing Strands in UHPC," *Proceedings of the 3rd International Symposium on High Performance Concrete/PCI National Bridge Conference*, October 19–22, 2003, Orlando, FL, Compact Disc, Paper 12.

Steinberg, E. and Reeves, E., "Structural Reliability of UHPC Bridge Girders in Flexure," *Proceedings of the PCI National Bridge Conference*, October 17–20, 2004, Atlanta, GA, Compact Disc, Paper 29.

Steinberg, E., "Structural Reliability of Prestressed UHPC Flexure Models for Bridge Girders," *Journal of Bridge Engineering*, Vol. 15, No. 1, 2010, pp. 65–72.

Stengel, T. and Schießl, P., "Life Cycle Assessment of UHPC Bridge Constructions: Sherbrooke Footbridge, Kassel Gätrnerplatz Footbridge and Wapello Road Bridge," *Architecture Civil Engineering Environment*, Vol. 1, 2009, pp. 109–118.

Stengel, T. and Schießl, P., "Sustainable Concrete with UHPC—From Life Cycle Inventory Data Collection to Environmental Impact Assessment," *Proceedings of the Second International Symposium on Ultra High Performance Concrete*, Ed., Fehling, E., Schmidt, M., and StLirwald. S., Kassel University Press, Kassel, Germany, 2008, pp. 461–468.

Stengel, T., "Sustainability Aspects of Traffic Bridges Made From UHPFRC—State-of-the-Art and Challenges for Concrete Technology," *Proceedings of the International Workshop on Ultra High Performance Fibre Reinforced Concrete—Designing and Building with UHPFRC: State of the Art Development*, 17–19 November 2009, Marseille, France, AFGC/*fib*, Paper 6.2.4.

Stiel, T., Karihaloo, B., and Fehling, E., "Effect of Casting Direction on the Mechanical Properties of CARDIFRC®," *Proceedings of the International Symposium on Ultra High Performance Concrete*, Ed., Schmidt, M., Fehling, E., and GeisenhanslLike, C., Kassel University Press, Kassel, Germany, 2004, pp. 481–493.

StLirwald, S. and Fehling, E., "Design of Reinforced UHPFRC in Flexure," *Proceedings of Hipermat 2012 3rd International Symposium on UHPC and Nanotechnology for High Performance Construction Materials*, Ed., Schmidt, M., Fehling, E., Glotzbach, C., Fröhlich, S., and Piotrowski, S., Kassel University Press, Kassel, Germany, 2012, pp. 443–450.

Sugamata, T. et al., "Characteristics of Concrete Containing a Shrinkage-Reducing Superplasticizer for Ultra-High-Strength Concrete," *8th CanMET/ACI International Conference on Superplasticizers and Other Chemical Admixtures in Concrete,* Publication No. SP-239, Ed., Malhotra, V.M., American Concrete Institute, Farmington Hills, MI, 2006, pp. 51–66.

Sugano, S., Kimura, H., and Shirai, K., "Experimental Studies on Seismic Behaviour of Columns and Interior Beam-Column Joints Which Used 200 MPa Fiber-Reinforced Concrete," *Proceedings of the Eighth U.S. National Conference on Earthquake Engineering*, 18–22 April 2006, San Francisco, CA, Paper 555.

Sujivorakul, C., "Flexural Model of Doubly Reinforced Concrete Beams Using Ultra High Performance Fiber Reinforced Concrete," *Proceedings of Hipermat 2012 3rd International Symposium on UHPC and Nanotechnology for High Performance Construction Materials*, Ed., Schmidt, M., Fehling, E., Glotzbach, C., Fröhlich, S., and Piotrowski, S., Kassel University Press, Kassel, Germany, 2012, pp. 435–442.

Suzuki, M., Maruyama, I., and Sato, R., "Properties of Expansive-Ultra High-Strength Concrete," *Seventh International Symposium on the Utilization of High-Strength/High-Performance Concrete*, Vol. II, Publication No. SP-228, Ed., Russell, H.G., American Concrete Institute, Farmington Hills, MI, 2005, pp. 1,159–1,173.

Swenty, M. and Graybeal, B., "Influence of Differential Deflection on Staged Construction Deck-Level Connections," FHWA, U.S. Department of Transportation, Report No. FHWA-HRT-12-057, National Technical Information Service Accession No. PB2012-111528, 2012.

Swenty, M. and Graybeal, B., "Material Characterization of Field-Cast Connection Grouts," FHWA, U.S. Department of Transportation, Report No. FHWA-HRT-13-041, National Technical Information Service Accession No. PB2013-130231, 2013.

Takatsu, H., "Experimental Study of Steel Fiber-Reinforced Ultra High-Strength Concrete Columns," Takenaka Technical Research Report 58, 2002, pp. 9–18. (in Chinese)

Talebinejad, I. et al. "Optimizing Mix Proportions of Normal Weight Reactive Powder Concrete with Strengths of 200–350 MPa," *Proceedings of the International Symposium on Ultra High Performance Concrete*, Ed., Schmidt, M., Fehling, E., and GeisenhanslLike, C., Kassel University Press, Kassel, Germany, 2004, pp. 133–141.

Tanaka, Y. et al., "Innovation and Application of UFC Bridges in Japan," *Proceedings of the International Workshop on Ultra High Performance Fibre Reinforced Concrete—Designing and Building with UHPFRC: State of the Art Development*, 17–19 November 2009, Marseille, France, AFGC/*fib*, Paper 3.1.4.

Tanaka, Y. et al., "Design and Construction of Sakata-Mirai Footbridge Using Reactive Powder Concrete," *Proceedings of the First fib Congress (FIB 2002)*, Osaka, Japan, Session 1—Big Projects and Innovative Structure, 2002, pp. 417–424.

Tanaka, Y. et al., "Technical Development of a Long Span Monorail Girder Applying Ultra High Strength Fiber Reinforced Concrete," *Proceedings of the Second International*

Symposium on Ultra High Performance Concrete, Ed., Fehling, E., Schmidt, M., and StLirwald. S., Kassel University Press, Kassel, Germany, 2008, pp. 803–810.

Tanesi, J., Graybeal, B., and Simon, M., "Effects of Curing Procedure on Freeze-Thaw Durability of Ultra-High Performance Concrete," *Proceedings of the 6th International RILEM Symposium on Fibre Reinforced Concretes (BEFIB' 2004)*, Publication pro039, Ed., di Prisco, M., Felicetti, R., and Plizzari, G., 2004, pp. 603–613.

Tavakoli, F., Bouteille, S., and Toutlemonde, F., "Application du Projet de Dalle Gaufrée du PN MIKTI sur un Ouvrage à Livron-Loriol," (Application of the MIKTI UHPC Waffle Deck Concept for a Bridge at Livron-Loriol), *Proceedings of the International Workshop on Ultra High Performance Fibre Reinforced Concrete—Designing and Building with UHPFRC: State of the Art Development*, 17–19 November 2009, Marseille, France, AFGC/*fib*, Paper 3.2.4.

Teichmann, T. and Schmidt, M., "Influence of the Packing Density of Fine Particles on Structure, Strength and Durability of UHPC," *Proceedings of the International Symposium on Ultra High Performance Concrete*, Ed., Schmidt, M., Fehling, E., and GeisenhanslLike, C., Kassel University Press, Kassel, Germany, 2004, pp. 313–323.

Teichmann, T. and Schmidt, M., "Mix Design and Durability of Ultra High Performance Concrete (UHPC)," *Proceedings of the 4th International Ph.D. Symposium in Civil Engineering*, Munich, Germany, 2002, pp. 341–347.

Teutsch, M. and Grunert, J., "Bending Design of Steel-Fibre-Strengthened UHPC," *Proceedings of the International Symposium on Ultra High Performance Concrete*, Ed., Schmidt, M., Fehling, E., and GeisenhanslLike, C., Kassel University Press, Kassel, Germany, 2004, pp. 523–532.

Thibaux, T. and Tanner, J. A., "Construction of the First French Road Bridges in Ultra High Performance Concrete," *Proceedings of the 1st fib Congress: Concrete Structures in the 21st Century*, Osaka, Japan, 2002.

Thibaux, T., "Renforcement Structurel de Bâtiments et d'Ouvrages de Génie Civil au Moyen de BFUP," (UHPFRC Applications in Structural Strengthening of Buildings and Civil Structures), *Proceedings of the International Workshop on Ultra High Performance Fibre Reinforced Concrete—Designing and Building with UHPFRC: State of the Art Development*, 17–19 November 2009, Marseille, France, AFGC/*fib*, Paper 5.1.3.

Thomas, M. et al., "Marine Performance of UHPC at Treat Island," *Proceedings of Hipermat 2012 3rd International Symposium on UHPC and Nanotechnology for High Performance Construction Materials*, Ed., Schmidt, M., Fehling, E., Glotzbach, C., Fröhlich, S., and Piotrowski, S., Kassel University Press, Kassel, Germany, 2012, pp. 365–370.

Toutlemonde, F. et al., "Efforts et Acquis de Recherche sur les BFUP depuis 2002 en vue de l'Actualisation des Recommandations de l'AFGC," (An Overview of Research Advances From 2002 Concerning UHPFRC, in View of Updating AFGC Recommendations), *Proceedings of the International Workshop on Ultra High Performance Fibre Reinforced Concrete—Designing and Building with UHPFRC: State of the Art Development*, 17–19 November 2009, Marseille, France, AFGC/*fib*, Paper 5.3.4.

Toutlemonde, F. et al., "Experimental Validation of a Ribbed UHPFRC Bridge Deck," *Proceedings of the Second International Symposium on Ultra High Performance Concrete*, Ed., Fehling, E., Schmidt, M., and StLirwald. S., Kassel University Press, Kassel, Germany, 2008, pp. 771–778.

Toutlemonde, F. et al., "Innovative Design of Ultra High-Performance Fibre Reinforced Concrete Ribbed Slab: Experimental Validation and Preliminary Detailed Analyses," *Seventh International Symposium on the Utilization of High-Strength/High-Performance Concrete*, Vol. II, Publication No. SP-228, Ed., Russell, H.G., American Concrete Institute, Farmington Hills, MI, 2005, pp. 1,187–1,205.

Tue, N.V. and KLichler, M., "Load and Deformation Behaviour of Confined Ultra High Performance Concrete Dowels," *Proceedings of the Second International Symposium on Ultra High Performance Concrete*, Ed., Fehling, E., Schmidt, M., and StLirwald. S., Kassel University Press, Kassel, Germany, 2008, pp. 553–560.

Tue, N.V. et al., "Application of UHPC Filled Tubes in Buildings and Bridges," *Proceedings of the International Symposium on Ultra High Performance Concrete*, Ed., Schmidt, M., Fehling, E., and GeisenhanslLike, C., Kassel University Press, Kassel, Germany, 2004, pp. 807–817.

Tue, N.V. et al., "Bearing Capacity of Stub Columns Made of NSC, HSC and UHPC Confined by a Steel Tube," *Proceedings of the International Symposium on Ultra High Performance Concrete*, Ed., Schmidt, M., Fehling, E., and GeisenhanslLike, C., Kassel University Press, Kassel, Germany, 2004, pp. 339–350.

Tue. N.V., Winkler, M., and Freytag, B., "Connections for Precast UHPC Parts," *Concrete Plant International*, North American Edition, No. 4, August 2011, pp. 50–56.

Turmo, J., Ramos, G., and Aparicio, A.C., "Shear Strength of Dry Joints of Concrete Panels With and Without Steel Fibres," *Engineering Structures,* No. 28, 2006, pp. 23–33.

Uzawa, T. et al., "Evaluation of Structural Performance of Ultra High Performance Concrete," *Proceedings of the First FIB Congress*, Osaka, 2002, pp. 77–82.

Van, V.T.A. and Ludwig, H.-M., "Proportioning Optimization of UHPC Containing Rice Husk Ash and Ground Granulated Blast-Furnace Slag," *Proceedings of Hipermat 2012 3rd International Symposium on UHPC and Nanotechnology for High Performance Construction Materials*, Ed., Schmidt, M., Fehling, E., Glotzbach, C., Fröhlich, S., and Piotrowski, S., Kassel University Press, Kassel, Germany, 2012, pp. 197–205.

Vande Voort, T.L., Sritharan, S., and Suleiman, M., "A Precast UHPC Pile for Substructural Applications," *Proceedings of the PCI National Bridge Conference*, October 22–24, 2007, Phoenix, AZ, Compact Disc, Paper 86.

Vande Voort, T., Suleiman, M., and Sritharan, S. "Design and Performance Verification of UHPC Piles for Deep Foundations," Final Report, Iowa Highway Research Board Project TR-558, Iowa State University, Ames, IA, 2008.

Vernet, C.P., "UHPC Microstructure and Related Durability Performance: Laboratory Assessment and Field Experience Examples," *Proceedings of the 3rd International Symposium on High Performance Concrete/PCI National Bridge Conference*, October 19–22, 2003, Orlando, FL, Compact Disc, Paper 31.

Visage, E.T. et al., "Experimental and Analytical Analysis of the Flexural Behavior of UHPC Beams," *Proceedings of Hipermat 2012 3rd International Symposium on UHPC and Nanotechnology for High Performance Construction Materials*, Ed., Schmidt, M., Fehling, E., Glotzbach, C., Fröhlich, S., and Piotrowski, S., Kassel University Press, Kassel, Germany, 2012, pp. 403–410.

Voo, J.Y.L., Foster, S.J., and Gilbert, R.I., "Shear Strength of Fibre Reinforced Reactive Powder Concrete Girders Without Stirrups," UNICIV Report No. R-421, The University of New South Wales, Studies from School of Civil and Environmental Engineering, Sydney, Australia, 2003.

Voo, Y.L. and Foster, S.J., "Malaysia First Ultra-High Performance Concrete Prestressed Motorway Bridge: Experimental Verification," 5th International Specialty Conference on Fiber Reinforced Materials, Singapore, August 28–29, 2008. Available at http://www.ju.edu.jo/sites/Academic/hunaiti/Lists/Published%20Research/Attachments/6/Abstract. pdf [Cited March 24, 2012].

Voo, Y.L., Foster, S.J., and Gilbert, R.I., "Shear Strength of Fiber Reinforced Reactive Powder Concrete Prestressed Girders without Stirrups," *Journal of Advanced Concrete Technology*, Vol. 4, No. 1, 2006, pp. 123–132.

Voo, Y.L., Poon, W.K., and Foster, S., "Shear Strength of Steel Fiber-Reinforced Ultrahigh-Performance Concrete Beams Without Stirrups," *Journal of Structural Engineering*, Vol. 136, No. 11, 2010, pp. 1,393–1,400.

Walraven, J.C., "Designing With Ultra High Strength Concrete: Basics, Potential and Perspectives," *Proceedings of the International Symposium on Ultra High Performance Concrete*, Ed., Schmidt, M., Fehling, E., and GeisenhanslLike, C., Kassel University Press, Kassel, Germany, 2004, pp. 853–864.

Walraven, J., "High Performance Fiber Reinforced Concrete: Progress in Knowledge and Design Codes," *Materials and Structures*, Vol. 42, No. 9, 2009, pp. 1,247–1,260.

Walraven, J., "On the Way to Design Recommendations for UHPFRC," *Proceedings of the Second International Symposium on Ultra High Performance Concrete*, Ed., Fehling, E., Schmidt, M., and StLirwald. S., Kassel University Press, Kassel, Germany, 2008, pp. 45–56.

Walraven, J., "On the Way to International Design Recommendations for Ultra High Performance Fibre Reinforced Concrete," *Proceedings of Hipermat 2012 3rd International Symposium on UHPC and Nanotechnology for High Performance Construction Materials*, Ed., Schmidt, M., Fehling, E., Glotzbach, C., Fröhlich, S., and Piotrowski, S., Kassel University Press, Kassel, Germany, 2012, pp. 51–58.

Walsh, K.K. and Steinberg E.P., "Moment Redistribution Capacity in Ultra-High Performance Concrete," *Proceedings of Hipermat 2012 3rd International Symposium on UHPC and Nanotechnology for High Performance Construction Materials*, Ed., Schmidt, M., Fehling, E., Glotzbach, C., Fröhlich, S., and Piotrowski, S., Kassel University Press, Kassel, Germany, 2012, pp. 639–646.

Wang, Y., Li, V.C., and Backer, S., "Experimental Determination of Tensile Behavior of Fiber Reinforced Concrete," *ACI Materials Journal*, Vol. 87, No. 5, September–October 1990, pp. 461–468.

Way, R. and Wille, K., "Material Characterization of an Ultra High-Performance-Fibre Reinforced Concrete Under Elevated Temperatures," *Proceedings of Hipermat 2012 3rd International Symposium on UHPC and Nanotechnology for High Performance Construction Materials*, Ed., Schmidt, M., Fehling, E., Glotzbach, C., Fröhlich, S., and Piotrowski, S., Kassel University Press, Kassel, Germany, 2012, pp. 565–572.

Wiens, U. and Schmidt, M., "State of the Art Report on Ultra High Performance Concrete of the German Committee for Structural Concrete (DAfStb)," *Proceedings of the Second International Symposium on Ultra High Performance Concrete*, Ed., Fehling, E., Schmidt, M., and Stürwald. S., Kassel University Press, Kassel, Germany, 2008, pp. 629–637.

Wille, K. and Parra-Montesinos, G., "Effect of Beam Size, Casting Method, and Support Conditions on Flexural Behavior of Ultra-High-Performance Fiber-Reinforced Concrete," *ACI Materials Journal*, Vol. 109, No. 3, May–June 2012, pp. 379–388.

Wille, K. and Naaman, A.E., "Pullout Behavior of High-Strength Steel Fibers Embedded in UltraHigh-Performance Concrete," *ACI Materials Journal*, Vol. 109, No. 4, July–August 2012, pp. 479–487.

Wille, K., Kim, D., and Naaman, A.E., "Strain-Hardening UHP-FRC With Low Fiber Contents," *Materials and Structures*, Vol. 44, No. 3, 2011, pp. 583–598.

Wille, K., Naaman, A.E., and El-Tawil, S., "Optimizing Ultra-High-Performance Fiber-Reinforced Concrete," *Concrete International,* Vol. 33, No. 9, September 2011, pp. 35–41.

Wille, K., Naaman, A.E., and Parra-Montesinos, G.J., "Ultra-High Performance Concrete With Compressive Strength Exceeding 150 MPa (22 ksi): A Simpler Way," *ACI Materials Journal*, Vol. 108, No. 1, January–February 2011, pp. 46–54.

Williams, E.M. et al., "Laboratory Characterization of Cor-Tuf Concrete With and Without Steel Fibers," Technical Report ERDO/GSL TR-09-22, U. S. Army Corps of Engineers, Engineer Research and Development Center, Washington, DC, 2009.

Wingenfeld, D., Muehlbauer, C., and Zilch, K., "Structural Behaviour and Load-Bearing Capacity of Reinforced Glued Joints of UHPC-Elements," *Proceedings of Hipermat 2012 3rd International Symposium on UHPC and Nanotechnology for High Performance Construction Materials*, Ed., Schmidt, M., Fehling, E., Glotzbach, C., Fröhlich, S., and Piotrowski, S., Kassel University Press, Kassel, Germany, 2012, pp. 709–716.

Wipf, T. et al., "Iowa's Ultra-High Performance Concrete Implementation," *Iowa Department of Transportation Research News*, April 2011.

Workshop on High Performance Fiber Reinforced Cementitious Composites in Structural Applications, Ed., Fischer, G. and Li, V.C., pp. 343–351.

Wu, C. et al., "Blast Testing of Ultra High Performance Fibre and FRP-Retrofitted Concrete Slabs," *Engineering Structures*, Vol. 31, No. 9, 2009, pp. 2060–2069.

Wu, X. and Han, S.-M., "First Diagonal Cracking and Ultimate Shear of I-Shaped Reinforced Girders of Ultra High Performance Fiber Reinforced Concrete Without Stirrup," *International Journal of Concrete Structures and Materials*, Vol. 3, No. 1, 2009, pp. 47–56.

Wu, X.-G. and Han, S.-M., "Interface Shear Connection Analysis of Ultrahigh-Performance Fiber-Reinforced Concrete Composite Girders," *Journal of Bridge Engineering*, Vol. 15, No. 5, 2010, pp. 493–502.

Xu, S., Hou, L., and Zhang, X., "Flexural Characteristics of Ultrahigh Toughness Cementitious Composite Beams with Different Depths," *Proceedings of the Third International fib Congress and Exhibition Incorporating the PCI Annual Convention and National Bridge Conference*, Washington, DC, May 29–June 2, 2010, Compact Disc, Paper 123.

Yan, G., Yan, Z., and Fang, Y., "An Optimization Design Procedure for Prestressed Reactive Powder Concrete Bridge," *Proceedings of the International Conference on Cement Combinations for Durable Concrete*, 5–7 July 2005, Dundee, Scotland, Ed., Dhir, R.K., Harrison, T.A., and Newlands, M.D., pp. 449–456.

Yan, P.Y. and Feng, J.W., "Mechanical Behavior of UHPC and UHPC Filled Steel Tubular Stub Columns," *Proceedings of the Second International Symposium on Ultra High*

Performance Concrete, Ed., Fehling, E., Schmidt, M., and Stürwald. S., Kassel University Press, Kassel, Germany, 2008, pp. 355–362.

Yang, H., Joh, C., and Kim, B.-S., "Structural Behavior of Ultra High Performance Concrete Beams Subjected to Bending," *Engineering Structures*, Vol. 32, 2010, pp. 3,478–3,487.

Yang, J. et al., "Characteristics of Mechanical Properties and Durability of Ultra-High Performance Concrete Incorporating Coarse Aggregate," *Proceedings of Hipermat 2012 3rd International Symposium on UHPC and Nanotechnology for High Performance Construction Materials*, Ed., Schmidt, M., Fehling, E., Glotzbach, C., Fröhlich, S., and Piotrowski, S., Kassel University Press, Kassel, Germany, 2012, pp. 257–264.

Ye, H.-W., "Technology of Ultra-High Strength/High Performance Concrete Applied to Guangzhou Xita Tower Construction," *Key Engineering Materials*, Vol. 405–406, 2009, pp. 1–4.

Young W.F. and Bopari, J., "Whiteman Creek Bridge—A Synthesis of Accelerated Bridge Construction (ABC), Ultra High Performance Concrete (UHPC), and Fiber Reinforced Polymer Concrete (FRP)," *Proceedings of the PCI National Bridge Conference*, October 22–26, 2011, Salt Lake City, UT, Compact Disc, Paper 111.

Young, W.F. et al., "Whitman Creek Bridge—A Synthesis of Ultra High Performance Concrete and Fiber Reinforced Polymers for Accelerated Bridge Construction," *Proceedings of Hipermat 2012 3rd International Symposium on UHPC and Nanotechnology for High Performance Construction Materials*, Ed., Schmidt, M., Fehling, E., Glotzbach, C., Fröhlich, S., and Piotrowski, S., Kassel University Press, Kassel, Germany, 2012, pp. 849–856.

Yuguang, Y., Walraven, J., and den Uijl, J., "Study on Bending Behavior of an UHPC Overlay on a Steel Orthotropic Deck," *Proceedings of the Second International Symposium on Ultra High Performance Concrete*, Ed., Fehling, E., Schmidt, M., and StLirwald. S., Kassel University Press, Kassel, Germany, 2008, pp. 639–646.

Zakariasen, D. and Perry, V., "Ultra-High Performance Concrete with Ductility: Design, Prototyping and Manufacturing of Panels and Boxes," *Thin Reinforced Cement-Based Products and Construction Systems*, Publication No. SP-224, Ed., Dubey, A., American Concrete Institute, Farmington Hills, MI, 2004, pp. 71–87.

Zehentner, H., "Innovative Precast Elements of High-Strength and Ultra-High-Strength Concrete to Increase the Punching Shear Resistance of Flat Slabs," BFT 73, 2007, (2), pp. 52–53, (in English and German).

Zhang, J., Stang, H., and Li, V., "Experimental Study on Crack Bridging in FRC Under Uniaxial Fatigue Tension," *Journal of Materials in Civil Engineering*, ASCE, Vol. 12, No. 1, February 2000, pp. 66–73.

Zheng, W., Kwan, A.K.H., and Lee, P.K.K., "Direct Tension Test of Concrete," *ACI Materials Journal*, Vol. 98, No. 1, January–February 2001, pp. 63–71.

Zimmermann, G. and Teichmann, T., "Membrane Concrete Grid Shells—UHPC Grid Shells," *Proceedings of the International Symposium on Ultra High Performance Concrete*, Ed., Schmidt, M., Fehling, E., and GeisenhanslLike, C., Kassel University Press, Kassel, Germany, 2004, pp. 839–852.

INDEX

A

B

C

D

E

F

G

H

I

U

V

W

Y